CHEMICAL ENGINEERING METHODS AND TECHNOLOGY

PLATINUM

COMPOUNDS, PRODUCTION AND APPLICATIONS

CHEMICAL ENGINEERING METHODS AND TECHNOLOGY

Additional books in this series can be found on Nova's website under the Series tab.

Additional e-books in this series can be found on Nova's website under the e-book tab.

MATERIALS SCIENCE AND TECHNOLOGIES

Additional books in this series can be found on Nova's website under the Series tab.

Additional e-books in this series can be found on Nova's website under the e-book tab.

CHEMICAL ENGINEERING METHODS AND TECHNOLOGY

PLATINUM

COMPOUNDS, PRODUCTION AND APPLICATIONS

LEONID VARENNIKOV
AND
EDUARD YEDEMSKY
EDITORS

nova publishers
New York

For permission to use material from this book please contact us:
Telephone 631-231-7269; Fax 631-231-8175
Web Site: http://www.novapublishers.com

NOTICE TO THE READER

LIBRARY OF CONGRESS CATALOGING-IN-PUBLICATION DATA

ISBN: 978-1-62257-939-6

Published by Nova Science Publishers, Inc. † New York

Contents

PREFACE

In this book, the authors present current research in the study of the compounds, production and applications of platinum. Topics discussed include the structural aspects of platinum coordination complexes; the problems of geochemical antagonism between REE and PGE in mafic and ultramafic rocks and their petrogenetic significance; the synthesis of platinum complexes of cyclic phosphines and bisphosphines; platinum films YSZ with different types of grain boundaries; platinum nanoparticles and inflammation; and platinum-halide square-prism compounds and their optical features.

Chapter 1 - Structural parameters of over two thousand and five hundred platinum coordination complexes are outline. There is wide variety of mer, from mono- to nanomer. In these platinum is found in the oxidation states of zero, +1, +2, +3, and +4, and also non-integral states of +0.5, +2.08, +2.14, +2.25, +2.5, +2.375, +2.75, +2.87, +3.25 and +3.5. There are also mixed-valence Pt(0)-Pt(II), Pt(I)-Pt(II), Pt(II)-Pt(III) and Pt(II)-Pt(IV). The most predominant state is Pt(II), with a square planar geometry and differing degree of distortion. In lower oxidation states (below 2), the Pt atoms are two, three and four coordinated, with the latter most prevails. The Pt(III) is mostly square-planar with some examples of trigonal bipyramidal arrangement. The Pt(IV) is mostly six-coordinated with various degree of distortion, there are examples of four-coordinated (square-planar). There is a wide variability of ligand types from uni- to hexadentate. The shortest Pt-Pt distance elongated in the order: 2.390(1) Å (Pt(III)-Pt(III)) < 2.492(1) Å (Pt(II)-Pt(II)) < 2.584(1) Å (Pt(2.5)-Pt(2.5)) < 2.619(4) Å (Pt(0)-Pt(0)) < 2.641(2) Å (Pt(0.5)-Pt(0.5)) < 2.699(1) Å (Pt(2.75)-Pt(2.75) < 2.724(2) Å (Pt(2.375)-Pt(2.375)) < 2.774(1) Å (Pt(2.25)-Pt(2.25)) < 2.824(1) Å (Pt(0)-Pt(II)) < 2.835(1) Å (Pt(2.08)-Pt(2.08)) < 2.839(2) Å (Pt(2.14)-Pt(2.14)) < 3.090(1) Å (Pt(IV)-Pt(IV)) < 3.331(1) Å (Pt(III)-Pt(IV)). Correlations between ligand donor atoms, bond lengths and bond angles are indicated and discussed.

Chapter 2 – It is known that REE began to be used as geochemical indicators for the diagnosis and classification of magmatic, metamorphic and metasomatic rocks and their minerals, as well as for reconstructions of the processes of their formation as developed the analytical methods for determining its quite low concentrations in the analyzed samples. At present, the analysis of REE in geological objects became a rather routine procedure even while studying the ultramafic and mafic rocks and minerals depleted by them. Due to this reason, the rare contemporary publications on the problems of Geochemistry, Mineralogy, Petrology of magmatic rocks and geodynamics do not use the results of REE analysis by one of the existing analytical methods. Approximately the same trend is observed with respect to platinum group elements (PGE), the determination of relatively low concentrations of which became possible with the introduction of several new analytical techniques including ICP-MS method.

In recent years, some studies were executed aiming to identify relationships between REE and PGE in rocks of ultramafic and mafic composition. Despite the limited number of analyses performed, the results of these studies have shown that comprehensive data on the contents of REE and PGE provide an opportunity to get some additional information that can be used in carrying out geochemical discrimination and systematization of ultramafic and mafic rocks, as well as massifs and complexes composed of them.

In this paper, the general aspects of relations between REE and PGE are discussed on the basis of the geological materials and more then 200 analyses for REE and PGE that were obtained on the 22 mafic-ultramafic massifs and complexes of various geneses that are located in regions of Russia, Mongolia, Kazakhstan, China, Germany, Greece, Vietnam, South Africa, Canada, Australia, Colombia, and others.

Chapter 3 – Novel platinum complexes incorporating phosphine ligands have been developed by us. On the one hand, platinum complexes of PtL_2Cl_2 type were synthesized, where L represents a 3-phospholene, phospholane, 1,2,3,6-tetrahydrophosphinine or 1,4-dihydrophosphinine moiety in differently substituted forms. In a few cases, the P-ligands were used in an optically active form. In the second group, 3-diphenylphosphino-substituted P-heterocycles, such as 3-Ph_2P-1-phenylphospholane, 3-Ph_2P-1-phenyl-1,2,3,6-tetrahydrophosphinine and 3-Ph_2P-1-phenyl-1,2,3,4,5,6-hexahydrophosphinine bisphosphines were synthesized to make available the corresponding *cis* chelate (ring) platinum complexes. Finally, bis(diphenyl phospino methyl)amines were utilized in the synthesis of other kind of ring platinum complexes. In the above cases, the phosphines and diphosphines were obtained from the corresponding phosphine oxide and bis(phosphine oxide) precursors after (double) deoxygenation by silanes. The cyclic bis(phosphine oxides) were obtained by the phospha-Michael reaction, while the bis(phosphinoxido methyl)amines by the double Kabachnik–Fields (phospha-Mannich) reaction. The new platinum complexes are potential catalysts in homogenous catalytic processes. Five platinum complexes were calculated by DFT and semiempirical (PM6-DH2) methods.

Chapter 4 – Platinum films were prepared on (111) and (100) orientated and on polycrystalline yttrium-stabilized zirconia (YSZ) by pulsed laser deposition (PLD). Afterwards, the Pt/YSZ systems were first annealed and then electrically polarised. The polarisation was performed in the manner that oxygen was alternatingly built-out and built-in (=oxygen exchange) at the Pt/YSZ. The as-deposited, annealed and polarised films, respectively, were characterised by EBSD, SEM, X-ray pattern and pole figure measurements. The as-deposited platinum films on all used substrates are (111) orientated, but the microstructures are different and depend strongly on the microstructure of the substrate. In as-deposited platinum films on (111) orientated YSZ the only defects are twins of a size between 100 nm and 300 nm. These twins disappear completely during annealing in platinum films on perfect (111) orientated YSZ single crystals and holes are built instead. Whereas on twin-rich (111) YSZ substrates at least a small fraction of twins remain. Nevertheless, during polarisation these twins also disappear completely, due to formation and cracking of oxygen bubbles at twin positions. The platinum films on (100) orientated YSZ are polycrystalline with mostly 30° and 60° grain boundaries. On polycrystalline YSZ the platinum films are formed with all kinds of grain boundary angles. Due to annealing the texture of the platinum films on (100) and polycrystalline YSZ becomes sharper, since the surface becomes smooth. Due to polarisation the grain boundaries are widened and the texture of the platinum films becomes softer, since the surface becomes waved. It can be concluded that twin grains within platinum films behave differently during annealing and oxygen exchange when compared to other grains. Presumably, this results from the much smaller disturbed regions within the crystal lattice at the twin grain boundaries.

Chapter 5 – Platinum nanoparticles (nano-Pts) possess anti-inflammatory activity and the ability to scavenge superoxides anions and peroxides, indicating that they can act as superoxide dismutase (SOD)/catalase mimetics. These potentials seem useful in the protection and/or the amelioration of oxidative stress-associated pathologies. Recent studies investigated the effects of poly acrylic acid-capped nano-Pts on ultraviolet (UV)-induced inflammation and the underlying molecular mechanisms in keratinocytes. Nano-Pts effectively protect against UV-induced inflammation in keratinocytes by decreasing reactive oxygen species (ROS) production. In addition, nano-Pts significantly inhibit RANKL-stimulated osteoclastogenesis in RAW 264.7 cells by the suppression of ROS. Moreover, the intranasal administration of nano-Pts prior to cigarette smoke exposure has been shown to inhibit inflammation in the lungs of mice. On the other hand, nano-Pts resulted in hyperthermia-desensitization in lymphoma cells where the signaling pathways involved in apoptosis are inhibited by nano-Pts. This chapter reviews the latest findings regarding the effects of nano-Pts on inflammatory responses.

Chapter 6 – Quasi-one-dimensional platinum-halide compounds are known by their unique optical properties and great effort is still devoted to fabricating their new varieties. Motivated by the newly designed square-prism-shaped iodo platinum complexes of general formula $[Pt(C_2H_8N_2)(C_{10}H_8N_2)I]_4(NO_3)_8\,nH_2O$, the authors construct a microscopic theory of their electronic structure with particular emphasis on the symmetry properties and in consideration of the experimental observations.

The authors reveal a variety of charge-density-wave states and illuminate their optical features.

In: Platinum: Compounds, Production and Applications
Editors: L.Varennikov and E. Yedemsky
ISBN: 978-1-62257-939-6

Chapter 1

STRUCTURAL ASPECTS OF PLATINUM COORDINATION COMPLEXES

M. Melník and P. Mikuš

Department of Pharmaceutical Analysis and Nuclear Pharmacy, Faculty of Pharmacy, Comenius University in Bratislava, Bratislava, Slovak Republic

ABSTRACT

Structural parameters of over two thousand and five hundred platinum coordination complexes are outline. There is wide variety of mer, from mono- to nanomer. In these platinum is found in the oxidation states of zero, +1, +2, +3, and +4, and also non-integral states of +0.5, +2.08, +2.14, +2.25, +2.5, +2.375, +2.75, +2.87, +3.25 and +3.5. There are also mixed-valence Pt(0)-Pt(II), Pt(I)-Pt(II), Pt(II)-Pt(III) and Pt(II)-Pt(IV). The most predominant state is Pt(II), with a square planar geometry and differing degree of distortion. In lower oxidation states (below 2), the Pt atoms are two, three and four coordinated, with the latter most prevails. The Pt(III) is mostly square-planar with some examples of trigonal bipyramidal arrangement. The Pt(IV) is mostly six-coordinated with various degree of distortion, there are examples of four-coordinated (square-planar). There is a wide variability of ligand types from uni- to hexadentate. The shortest Pt-Pt distance elongated in the order: 2.390(1) Å (Pt(III)-Pt(III)) < 2.492(1) Å (Pt(II)-Pt(II)) < 2.584(1) Å (Pt(2.5)-Pt(2.5)) < 2.619(4) Å (Pt(0)-Pt(0)) < 2.641(2) Å (Pt(0.5)-Pt(0.5)) < 2.699(1) Å (Pt(2.75)-Pt(2.75) < 2.724(2) Å (Pt(2.375)-Pt(2.375)) < 2.774(1) Å (Pt(2.25)-Pt(2.25)) < 2.824(1) Å (Pt(0)-Pt(II)) < 2.835(1) Å (Pt(2.08)-Pt(2.08)) < 2.839(2) Å (Pt(2.14)-Pt(2.14)) < 3.090(1) Å (Pt(IV)-Pt(IV)) < 3.331(1) Å (Pt(III)-Pt(IV)). Correlations between ligand donor atoms, bond lengths and bond angles are indicated and discussed.

1. INTRODUCTION

Platinum is one of the most extensive uses as a catalyst in a wide variety of processes, and of course it continues use as a jeweler metal. It should not be overlooked that platinum has played a crucial role in the development of many branches of science. Reliable platinum crucibles were vital in classical analysis which laid the foundations of modern chemistry.

Platinum exists in a wide range of oxidation state from zero to six, of which +2 and +4 are the most common, particularly in coordination chemistry. Platinum(II) has a strong preference for square planar geometry, of which the kinetic inertness has allowed extensive use of these complexes in the characterization of cis-trans isomerism and reaction mechanisms. Much work in this field has come from the Russian school such as the contributions of Chernyaev. Already in the 1920s he noticed that when there are alternative positions at which an incoming ligand might effect a substitution, the position chosen depends not so much on the two ligands directly involved as on the nature of the inactive ligand trans to that position. This became known as the „trans effect“ and has had a considerable influence on the synthetic chemistry of Pt(II).

A resurgence of interest in platinum complexes started in 1969 when Rosenberg et al. discovered the anti-tumor activity of $Pt(NH_3)_2Cl_2$ known as „cisplatin“ [1]. These stimulate considerable research activity into the chemistry of platinum which has resulted in the determination of a considerable number of structures by diffraction techniques. Since structure is often related to chemical and biological activity for such reason we analyzed and classified crystallographic and structural parameters for platinum coordination complexes as well as platinum borates. A comprehensive overview on such data and their detail descriptions for over two thousand and five hundred examples were done by us [2-7]. Therefore in this Chapter we will outline only briefly view on the structures for such reason that there are over two thousand references we turn an attention on the review articles and cited references therein, as well.

2. Monomeric Pt Coordination Complexes

2.1. Pt(0) and Pt(I) Complexes

There are thirty derivatives of monomeric Pt(0) coordination complexes for which structural parameters are available. These complexes crystallized in monoclinic (x16), triclinic (x9), orthorhombic (x2), rhombohedral, tetragonal and hexagonal each (x1). The Pt(0) is two- PtP_2, three- PtP_3 and four- PtP_4, PtP_2S, PtO_2P_2, PtN_2P_2, PtS_2P, PtP_2ON, PtP_2NS and Pt_2CS, coordinated. Two P atoms in PtP_2 has the P-Pt-P angles range from 160.5(1) to 180.0°, with the mean Pt(0)-P bond distance of 2.244 Å. The PtP_3 is a Y-shape of structure with the mean Pt0)-P bond distance of 2.278 Å. The mean deviation of P-Pt-P bond angles from the ideal 120° is 3.8° (range 2.7 to 5.0°). Mostly four coordinated Pt(0) derivatives are tetrahedral with differing degree of distortion. The mean Pt(0)-L bond distance elongated in the order: 2.102 Å (bi-O_2) < 2.282 Å (PL, mostly PPh_3) < 2.310Å (bi-PL) < 2.372Å (SL). The bidentate chelated ligands created n-membered metallocyclic rings, which opens in the order: 42.9°(-OO-) < 52.8°(-SP-) < 53°(-PP-) < 96.4°(-PC_3P-) < 115.3°(-PC_6P-). The Pt(I) complexes are four-coordinated with inner coordination spheres PtN_4 and PtP_4. The mean Pt-L bond distance elongated in the order: 2.03 Å (bi-NL) < 2.29 Å (PL) < 2.35 Å (bi-PL). The chelated ligands form five- N-Pt-N resp. four- P-Pt-P metallocycles with the values of 79.2 and 52.3°, respectively.

2.2. Pt(II) Complexes

There are over one thousand and two hundred monomeric Pt(II) coordination complexes for which crystallographic and structural parameters are available. The most common crystal classes are monoclinic and triclinic. Square planar complexes dominate with a few examples of a trigonal-bipyramidal environment. The inner coordination spheres in the series of a square-planar can be divided into the four main groups:

a) PtA_4 – with homo- and quasi –homogeneous coordination sphere
b) PtA_3B – with three homo- and quasi –homo- donor ligands and one is differing
c) PaA_2B_2 – with two pairs of homo- and quasi –homo- donor ligands
d) PtABCD – four variation types of donor atoms.

To the first group belongs: PtA_4 (A=O, N, C, Cl, S, P, Br, Se and As). The PtO_4 coordination is created only by bidentate chelated O-donor ligands. These chelated ligands are two types, ones which created five-membered and another one six-membered metallocycles. The mean Pt-O bond distances and O-Pt-O bond angles (five- vs six-) are 1.967(6) Å and 81.3(2)° vs 2.012(3) Å vs 94.8(6)°.

A compound with PtN4 chromophores by far prevails. These can be formally divided into two main groups:

(a) with a homogenous coordination sphere (all four N-donor ligands are identical);
(b) with a quasi- homogenous coordination sphere. This (b) group can be divided into four subgroups:
(b_1) three identical ligands plus one other- PtN_3N^I;
(b_2) two pairs of identical ligands- $PtN_2N_2^I$;
(b_3) one identical pair and two different ligands- $PtN_2N^IN^{II}$;
(b_4) all ligands are different- $PtN^IN^{II}N^{III}N^{IV}$.

The (a) group in most often created by four amine ligands, four alkyl N-donor ligands or aryl N-donor ligands. The mean Pt-N bond distance elongated in the order: 2.01 Å (N-alkyl) < 2.02 Å (N-aryl) < 2.055 Å (NH_3). There are only three examples belonging to the (b_1) subgroup. The mean Pt-N bond distance elongated in the order: 2.055 Å (NL) < 2.03 Å (NH_3). The most common type subgroup is actually (b_2), $PtN_2N_2^I$. Practically all examples are of the composition $\{Pt(NH_3)_2(LN^I)_2\}^{+2}$ with both cis- and trans- derivatives. The mean Pt-NH_3 bond distance in the cis- derivatives is 2.04 Å which is slightly shorter than in the trans- derivatives at 2.06 Å. For the composition $\{Pt(MeNH_2)_2(LN^I)_2\}^{+2}$ only trans derivatives are found with a mean Pt-N(H_2Me) bond distance of 2.06 Å. Another derivatives of the composition {Pt(alkyl amine)$_2$(LN)$_2$ (LN- arylamine or NO_2) are only cis- derivatives. Several other derivatives have the (b_3) type of composition $\{Pt(NH_3)_2N^IN^{II}\}^{+2}$ and again exhibit cis- as well as trans- derivatives. The mean Pt-NH_3 bond distances for both cis- and trans- derivatives are the same of 2.05Å. Finally, there is an example which belongs to the (b_4) subgroup and that is {Pt(NH_3)(1-methyl-cytosine)(9-methyladenine)(9-ethylquaninate)}$^+$. In the series of PtN_4 derivatives in which all N- donor ligands are unidentate the mean cis-

and trans- N-Pt-N bond angles deviated from the ideal angles of 90 and 180° of 1.9° (range 0 - 6.5°) and 3.45° (range 0 - 9°).

There are over thirty derivatives of the composition Pt(η^2-NL)(NL)$_2$. The most common bidentate chelate ligands are 2,2′-bipyridyl (bpy) and its derivatives for the unsaturated N-donor ligands, and ethylenediamine (en) and its derivatives for the saturated N-donor ligands. Practically all the complexes are cis- derivatives. The mean values of Pt-N(η^2) bond distance and N-Pt-N(η^2) bond angle in the Pt(η^2-bpy)(NL)$_2$ derivatives of 2.005 Å and 80.5° are smaller that those found in Pt(η^2-en)(NL)$_2$ with the values of 2.04Å and 84.0°. The mean Pt-N (unidentate) bond distance elongated in the sequence: 2.005 Å (NCR) < 2.015 Å (pyridine, pyrazole) < 2.02 Å (N-nucleoside) < 2.05 Å (NH$_3$).

There are almost eighty examples of the compositions Pt(η^2-NL)$_2$ and Pt(η^2-NL)(η^2-NL′) in which each Pt(II) has a square-planar environment (PtN$_4$) as well, with various degree of distortion. There are several types of metallocyclic rings which created by the bidentate chelate ligands. The mean Pt-N bond distance for saturated N-donor ligands of 2.035 Å is somewhat longer than that found for unsaturated N-donor ligands with the values of 1.995 Å. The metallocyclic rings (N-Pt-N) open in the order: 63.7° (-NCN-) < 78.5° (-NC$_2$N-, unsaturated) < 82.2° (-NC$_2$N-, saturated) < 89.7° (-NC$_3$N-) < 96.9° (-NC$_4$N-).

There are almost twenty Pt(η^3-NL)(NL) complexes in which Pt(II) atom has a square planar environment (PtN$_4$) with varying degree of distortion. There are actually two groups, one colored where the terdentate ligand is 2,2′: 6′,2″ -terpyridine (tpy) and other colorless where the terdentate ligand is diethylenetriamine (dien). The central N-atom of the terdentate ligands (tpp, dien) are somewhat close to the Pt(II) atom with the values of 1.94 (tpp) and 1.99 Å (dien), than the remaining two with the mean lengths of 2.03 and 2.05 Å, respectively. The five-membered metallocycles are smaller than ideal 90°, with the mean values 80.9° (tpp) and 84.7° (dien). The mean Pt-N (unidentate) bond distances are 2.027 and 2.032 Å, respectively.

There are over twenty Pt(η^4-NL) derivatives in which a single tetradentate ligand provides a square planar geometry around a Pt(II) atom (PtN$_4$). These derivatives are divided into two groups, one with three-metallocyclic rings and the other with four metallocyclic rings. In the first group a tetradentate ligands create three five-membered metallocyclic rings with the mean intra Pt-N bond distance of 1.985 Å and the mean inter Pt-N bond distance of 2.045 Å. The N-Pt-N bond angles range from 78.5(3) to 88.9(4)° with the average of 82.8°. Another derivatives have a middle six-membered metallocycle and two outer five-member chelate rings with the mean N-Pt-N values of 97.6(2) and 79.6(4)°, respectively. The mean Pt-N bond distances are 1.99Å for intra-N atoms and 2.01 Å for inter-N atoms. There is an example in which tetradentate corrinoid ligand makes three six-membered rings with N-Pt-N bond angles of 90° (central) and 92° (outer) and the mean Pt-N bond distances of 2.00 and 2.01 Å, respectively.

In the second group with four metallocyclic rings the most common are those which form two pairs of rings, five- and six-membered in trans position. The mean N-Pt-N value for the former is 84.3° (range 83 to 85°) and for the latter is 95.5° (range 95 to 96.4°), as expected. The mean Pt-N bond distance is 2.04 Å (range 1.97 to 2.10 Å). There are examples contain a 16 membered macrocycles which creates two pairs of metallocyclic rings, five- and seven-membered in trans positions, or four six-membered metallocyclic rings. The mean N-Pt-N angle opens in the order: 84.9(2)° (five-) < 90.9° (six-) < 95.1(2)° (seven-). The mean Pt-N bond distance is 2.00 Å.

There are almost thirty, mostly black Pt(II) derivatives where four CN groups are arranged in a square planar geometry around a Pt(II) atom PtC_4). The Pt-C bond distances range from 1.93 to 2.15 Å with an average value of 1.99 Å. The cis- and trans- C-Pt-C angles range from 86 to 94° (mean 90°) and from 176 to 180°, respectively. Also thirty derivatives $[PtCl_4]^{2-}$ were found. The cis-and trans- Cl-Pt-Cl bond angles depart from the ideal values of 90° and 180° by 1.9° and 2.9°, respectively. The mean Pt-Cl bond distance is 2.30 Å (range 2.23 – 2.32 Å).

There are more than sixty derivatives, mostly monoclinic (x39) and triclinic (x17) with some orthorhombic (x5) and two tetragonal. The square planar geometry is realized by mono-, bi- and tetradentate S-donor ligands (PtS_4). Thirteen derivatives have only monodentate ligands with the mean Pt-S bond distance of 2.32 Å (range 2.30 – 2.34 Å). The cis- and trans- S-Pt-S bond angles depart from the ideal (90, 180°) by 9° and 10°, respectively. There are derivatives with the composition $(Pt(\eta^2\text{-SL})(SL)_2$, $Pt(\eta^2\text{-SL})(\eta^2\text{-SL})$ (most common) and $Pt(\eta^4\text{-SL})$. The homo-bidentate chelated ligands created four-, five- and six-membered metallocycles. The mean Pt-S bond distances and S-Pt-S bond angles open in the orders: 2.32 Å, 75° (-SCS-) < 2.34 Å, 83.6° (-SPS-); 2.29 Å, 89.3° ($-SC_2S-$) < 2.32 Å, 93.4° (-SSSS-) < 2.24 Å, 93.6° (-SSCS-). The six-membered rings open in the order: 90.0° (-SPNPS-) < 92.7° (-SCNCS-) < 97.5° ($-SC_3S-$). In the same order the mean Pt-S bond distance shortened: 2.335 Å > 2.295 Å > 2.28 Å.

Tetradentate S-donor ligands created five-, six-, seven- and even eight-membered metallocyclic rings with the mean S-Pt-S "bite" angles open in the order: 86.3° (five-) < 90.6° (six-, eight-) < 93.6° (seven-). The mean Pt-S bond elongated in the order: 2.275 Å (five-) < 2.30 Å (six-) < 2.31 Å (eight-) < 2.405 Å (seven-).

There are forty PtP_4 derivatives; they are mostly monoclinic and colorless, with mono-, bi- and tetradentate ligands, bidentate being the most common. In $Pt(PL)_4$ derivatives, the mean Pt-P bond distance is 2.32 Å (range 2.29 – 2.37 Å). The deviations of cis- and trans- P-Pt-P bond angles from ideal (90 and 180°) are 4° and 13°. In $Pt(\eta^2\text{-PL})(PL)_2$ the four-membered rings with P-Pt-P bond angles opening in the sequence: 70.3° (-PCP-) < 74.7° (-PBP-) < 76.8° (-PPP-) and six-membered ring with PNCNP atoms has a P-Pt-P ring angle of 79.6°. In $Pt(\eta^2\text{-PL})(\eta^2\text{-PL})$ derivatives chelated ligands created with a Pt(II) several metallocycle types. The mean P-Pt-P "bite" angles open in the order: 68.7° (-PNP-) < 71.3° (-PCP-) < 85.0° (-PCOP) < 86.5° ($-PC_2P-$); 86.5° (-PNPNP-) < 88.5° (-PCHCP-) < 90.0° (-POBOP-) < 92.9° ($-PC_3P-$). The mean Pt-P bond distance is 2.32 Å.

There are derivatives with the following inner coordination spheres $PtBr_4$ (x2), $PtSe_4$ (x7), and $PtAs_4$ (x1). The mean Pt-L bond distance elongated in the order: 2.375 Å (di- AsL) < 2.42 Å (Br) < 2.425 Å (di- SeL) < 2.44 Å (SL). There are several types of metallocycles with the mean L-Pt-L bond angles open in the order: 77.8° (-SeCSe-) < 84.5° ($-AsC_2As-$) < 89.0° ($-SeC_3Se-$) < 99.4° ($-SeSe_2Se-$, -SeNPPSe-).

There are over one hundred and seventy Pt(II) derivatives with the following inner coordination spheres: PtO_3X (X=N, P, Br or I), PtN_3X (X=O, C, Cl, S, Br or I), $PtCl_3X$ (X=N or S), PtS_3X (X=O, N, Cl, P, As or I), PtP_3X (X=H, F, Cl, S, Br or I), $PtBr_3X$ (X=N or Si) and PtI_3X (X=P or Sb). The mean Pt-L bond distance for monodentate ligands elongated with the covalent radius of donor atom except SL, in the order: 1.75 Å (H, 0.37 Å) < 2.02 Å (OL, 0.73 Å) < 2.03 Å (N, 0.75 Å) < 2.33 Å (Cl, 0.99 Å) < 2.335 Å (SL, 1.02 Å) < 2.34 Å (PL, 1.06 Å) < 2.43 Å (Br, 1.13 Å) < 2.595 Å (I, 1.33 Å). For bi- and terdentate ligands the orders are: 2.04 Å (bi- NL) < 2.24 Å (bi- SL) < 2.28 Å (bi- PL); and 2.01 Å (ter- NL) < 2.27 Å (ter- PL) <

2.31 Å (ter- SL). There are hetero donor ligands as well O/S, S/P, N/O_2 and N/S_2. The bi- and terdentate ligands created several types of metallocycles with Pt(II), four-, five-, six- and even seven- membered rings. The L-Pt-L bond angles open in the order: 70.0° (-PNP-) < 71.5° (-PCP-) < 74.2° (-SCS-) < 78.2° (-PPP-) < 82.7° (-SPS-); 81.0° (-NC_2N-, unsaturated) < 85.7° (-SNCP-) <85.8° (-OC_2S-) ~ 85.8° (-NC_2S-, unsaturated); 92.4° (-PC_3P-) < 93.0° (-NC_3N-) < 93.5° (-SC_3S-), and 89.0° (-NC_4N-). The most frequent derivatives in the series of PtA_3B are PtN_3X (26%), $PtCl_3X$ (17%) and PtP_3X (11%).

There are over seven hundred and twenty Pt(II) complexes of the PtA_2B_2 composition. These complexes have the following inner coordination spheres: PtO_2X_2 (X=N, Cl, S, P or As), PtN_2X_2 (X=C, Cl, S, P, Br or I), PtC_2X_2 (X=Cl, P or I), PtB_2P_2, $PtCl_2X_2$ (X=S, P, Se or As), PtS_2X_2 (X=P, Br, As or I), PtP_2X_2 (X=Br, Se, I or Te), $PtBr_2X_2$ (X=Se or I), and PtI_2Te_2. Each Pt(II) has a square planar environment with differing degree of distortion. These complexes crystallized mainly as monoclinic and triclinic. The most common inner coordination spheres are PtN_2Cl_2 (25%), PtO_2N_2 (17.5%) and $PtCl_2P_2$ (17%). There is wide variety of coordinated ligands. The mean Pt-L (monodentate) bond distance elongated in the order: 1.55 Å (H) < 1.965 Å (CN) < 1.99 Å (NL, unsaturated) < 2.01 Å (OL) < 2.06 Å (NL, saturated; BL) < 2.28 Å (SL, PL) < 2.31 Å (Cl) < 2.365 Å (SiL) < 2.405 Å (SeL) < 2.415 Å (Br) < 2.59 Å (F). The mean Pt-L (bidentate) bond distance elongated in the order: 2.02 Å (2NL, unsaturated) < 2.045 Å (2NL, saturated) < 2.06 Å (2OL) < 2.255 Å (2SL) < 2.270 Å (2PL) < 2.375 Å (2SiL) < 2.383 Å (2AsL) < 2.435 Å (2 SeL) < 2.582 Å (2TeL). There is also wide variety of hetero- donor ligands with the donor sites: O/N, O/S, O/P, N/S, N/P, S/As, P/Te, P/Si, P/Se, O_2/N, O_2/P, N/S_2 and N_2/S_2. Wide variety of metallocycles is created by homo- as well as hetero-dentate ligands. Both electronic and steric factors appear in the variation of the internal L-Pt-L angles of metallocyclic rings. The L-Pt-L bond angle for the PtA_2B_2 type derivatives open in the sequences: 65° (OCO, SiOSi) < 71.5° (OPO) < 73° (PCP) < 74°(SCP) < 74.3° (OSeO) < 76.8° (SCS) < 77° (SeCSe) < 80.7° (SSS) < 81.7° (SPS) < 85.3° (SePSe) -4-membered rings; 76.4° (NN_2N) < 77° (PC_2Si) < 78° (ON_2O) < 79.5° (NC_2N, unsaturated) < 79.7° (NCPN) < 80.6° (OCNO) < 81.5° (OC_2N) < 81.7° (NCON) < 82.0° (SeC_2Se) < 82.2° (OC_2O) < 82.5° (NCSN) < 83.7° (OC_2P, NC_2N saturated) < 84.8° (NC_2P) < 85.0° (NC_2S) < 86.3° (PC_2P) < 87.4° (SC_2As) < 88.5° (PC_2Te, TeC_2Te) < 89.0° (SC_2S) < 89.4° (SSNS) < 89.9° (SS_2S) < 92.4° (PCNSe) -5-membered rings; 87.4° (SC_3P) < 88.7° (NSNSN) < 89.0° (NCPCN) < 89.5° (NCNPN) < 89.9° (SC_3S) < 90.5° (OC_3O) < 91.2° (NC_3N) < 94.5° (PC_3P) < 96.0° (SPNPS) < 98.4° (SSNPS) -6-membered rings; 88.5° (NC_4N, unsaturated) < 89.7° (OC_4O) < 91.6° (SC_4S) < 98.8° (PC_4P, NC_4N saturated) -7-membered rings; and 90.3° (OC_5O) < 102.2° (PC_5P) < 102.4° (SC_2SC_2S) -8-membered rings. Some derivatives contain makro-metallocycles with the L-Pt-L angles: 87.0° (SC_8S) < 99.6° (PC_6P) < 101.3° ($PC_{17}P$) < 174.0° (AsC_9As) < 174.4° ($PC_{12}P$) < 176.9° ($OC_{11}P$) < 178.3° (PC_9P). Clearly the conformation of the largest especially eleven (SC_8S), twenty ($PC_{17}P$)-membered rings are fairly rigid.

There are over two hundred and seventy Pt(II) derivatives of the composition PtA_2BC in which each Pt(II) atom is in a distorted square-planar arrangement. The inner coordination spheres about the Pt(II) are: PtO_2NX (X=Cl, P or I), PtO_2ClS, PtN_2OX (X=Cl or S), PtN_2ClX (X=S or P), PtN_2PX (X=S or Se), PtC_2NS, $PtCl_2OX$ (X=N, S or P), $PtCl_2NX$ (X=S, P, Se, As or Te), $PtCl_2SX$ (X=P or Te), PtP_2HX (X=O, N, Cl, S, Br, Si or I), PtP_2OX (X=N, Cl, S, Br or Si), PtP_2NX (X=Cl, S, Br, Se, I or Te), PtP_2BCl, PtP_2ClX (X=S, Si or Te), PtP_2BrSi, PtP_2SiTe, $PtBr_2NS$, $PtBr_2ClP$, $PtSe_2SlP$, $PtAs_2ClS$, PtI_2NX (X=S or P), PtI_2SP) and $PtTe_2NCl$.

The most frequent are PtP_2AB (55%). These complexes crystallized mostly in monoclinic and triclinic classes. The inner coordination spheres about the Pt(II) are created wide variety of mono-, bi- (homo- as well as hetero-), and hetero-terdentate ligands. The mean Pt-L (monodentate) bond distance elongated in the sequence: 1.57 Å (H) < 1.97 Å (NC) < 2.01 Å (BL) < 2.06 Å (OL) < 2.095 Å (NL) < 2.24 Å(SL) < 2.27 Å (PL) < 2.31 Å (Cl) < 2.345 Å (AsL) < 3.55 Å (SiL) < 2.43 Å (SeL) < 2.475 Å (Br) < 2.618 Å (I) < 2.655 Å (TeL). The mean Pt-L (homobidentate) bond distance elongated in the sequence: 1.975 Å (OL) < 1.985 Å (NL, unsaturated) < 2.055 Å (NL, saturated) < 2.245 Å (PL) < 2.31 Å (SL) < 2.375 Å (AsL, SiL) < 2.820 Å (SeL). The heterobidentate chelated ligands are with O+N, O+S, O+P, N+S, N+P, N+Se, N+Te, S+P, P+B and S+Te donor sites. The heteroterdentate ligands are with 2O+N, 2N+O, 2N+S, 2N+Se, 2S+N, 2P+Si and 2As+S donor sites. These chelated ligands created a wide variety of metallocyclic rings with a Pt(II), and the effects of both steric and electronic factors can be seen from the values of the L-Pt-L bond angles: 51.0° (NP) < 61.0° (OP); 70.3° (NCS) < 70.8° (NCP) < 73.5° (OCB) < 75.0° (SCS) < 78.0° (NCSe); 80.0° (NC_2N, unsaturated) < 81.7° (OC_2N) < 84.0° (OC_2S; NC_2P) < 84.7° (OC_2P) < 85.0° (NC_2N, saturated) < 85.5° (NC_2S) < 86.5° (PC_2P) < 87.4° (NC_2Se) < 87.8° (NC_2Te)< 88.7° (SC_2P) < 91.0° (SC_2S) < 95.5° (SeSeNSe); 87.0° (PC_3P) < 93.7° (OC_3N, NC_3Se) < 95.0° (NC_3S) < 95.3° (OC_3S) < 96.8° (SC_3S). There are also seven- 86.0° (NC_4N) < 102.9° (PC_4P);eight- 89.5° (NC_5N) and even twelve- 176.2° (PC_9P) membered metallocycles. The most common ligands are PPh_3, dmso (S) and chlorine.

There are thirteen Pt(II) examples, in which a distorted square-planar environment about a Pt(II) is created by four different donor atoms. In PtONClS (x5) and PtONClP (x4) derivatives a hetero-bidentate (O,N) created five- in seven derivatives and in the remaining two derivatives six-membered metallocycles are created with the mean O-Pt-N "bite" angles of 81.0° and 92.0°, respectively. The chlorine atoms with S (dmso) or P (PL) ligands completed an environments. Inner coordination sphere PtNClSP is forms by a terdentate-N,S,P ligand which created five- (N,S) and six- (N,P)- membered metallocycles, with the N-Pt-S and N-Pt-P "bite" angles of 84.9(4) and 95.8(4)°, respectively. Chlorine completed a distorted square-planar geometry. Inner coordination spheres PtNSPX (X=Br or I) are created by a heterobidentate- N(H)SNS ligand with PMe_2Ph and X atom. The chelate ligand forms a five metallocycle with the mean N-Pt-S "bite" angle of 85.5°. In one example an inner coordination sphere- PtNSPI is created by a terdentate- N,S,P ligand and a iodine atom. The terdentate ligand forms with Pt(II) two six-metallocycles with the N-Pt-P and N-Pt-S angles of 89.1 and 93.2°, respectively. The mean Pt-L (monodentate) bond distance elongated in the order: 2.185 Å (SL) < 2.305 Å (Cl) < 2.235 Å (PL) < 2.475 Å (Br) < 2.507 Å (I).

In the series of a square-planar Pt(II) complexes are five main types of the inner coordination sphere: PtA_4 (x455), PtA_3B (x175), PtA_2B_2 (x750); PtA_2BC (x280) and PtABCD (x15). There is a wide variety of uni-, bi- (both homo- and hetero-) and heteroterdentate ligands which form a square-planar coordination about the Pt(II) atom. Platinum in oxidation state +2 should be preferentially bonded to "soft" donor atoms, most commonly PPh_3 and S-donor ligands. However, Pt(II) can be found here bonded to both „borderline" and „hard" donor atoms, most commonly the Cl atom.

The mean Pt-L bond lengths for the soft and borderline ligands increase with the covalent radius of the donor atom. A similar trend is observed with the hard ligands, but it does not overlap the former. The mean Pt-L bond lengths for the homobidentate ligands are somewhat shorter than those of the corresponding monodentate ligands, with the exception of the P-donor ligands in which the opposite trend is seen. The homobidentate ligands include: O , N , S-, P-, Si-, Se-, As-, and Te. There is a wide variety of heterobidentate ligands: O + N, O + S, O + P; N + S, N + P, N + Se, N + Te; B + P; S + P, S + As, S + Te; P + Se, P + Si; and heteroterdentate ligands: 20 + N, 2N + 0, 2N + S, 2N + Se, 2S + N, 2P + Si and 2As + S.

On the distortion of square-planar geometry about Pt(II) atom influenced at least two factors, a combination of donor atoms/ligands and „size" of metallocycles. In the series when an inner coordination sphere is build up only by unidentate ligands, the mean deviation from the ideal L-Pt-L bond angles of 90°(cis-) and 180°(trans-) increases in the order: 2.4° (range 0-11°) and 4.4° (0-9.7°) {PtA_4, 175 examples} < 2.7 (0-9.7°) and 5.3° (0-12.6°) {PtA_2B_2. 90 examples} < 3.6° (0.3-12.8°) and 7.4° (0.7-23.2°) {PtA_3B, 175 examples}. This reflects that the distortion increases in the given order. It noticeable, that highest degree of distortion was found in the following types: PtP_4 < PtO_2S_2 < PtP_2HX < PtP_3H. The „size" of metallocycles as the second reason influenced on the degree of square-planar distortion as was mentioned that is wide variety of homo- as well as hetero- bi- and terdentate ligands with created from three- to nine-, eleven-, twelve-, fourteen-, sixteen- and even twenty-membered metallocycles. The mean L-P-L „bite" angles open in the order: 43.5° (three-) < 66.2° (four-) < 87.0° (eleven-) < 88° (five-) < 91.4° (six-) < 91.9° (seven-) < 95.0° (nine-) < 95.8° (eight-) < 99.0° (sixteen-, twenty-) < 152° (twelve-) < 175.8° (fourteen-) membered rings. Such values influence on the degree of distortion of the respective square-planar geometry as well.

There are only seven, mostly yellow pentacoordinated Pt(II) complexes, with the following inner coordination spheres: PtP_5, PtP_4H (3 examples), PtP_4N, $PtAs_4I$ and PtP_3As. Each Pt(II) atom has a trigonal bipyramidal environment. In one example a pair of homo-bidentate $Ph_2PCH_2CH_2PPh_2$ ligands with a hydride forms PtP_4H. One of the ligand coordinated in the equatorial side with the mean Pt-P bond distance of 2.322(2), and the other span an equatorial (Pt-P, 2.271(2) Å) and axial (2.408(2) Å) position. A hydride also bound in an axial position. In all remainders a tetradentate, $P(CH_2CH_2PPh_2)_3(P_4)$, $As(C_6H_4AsPh_2)_3(As_4)$, and $N(CH_2CH_2PPh_2)_3(P_3N)$ with unidentate H, PL, AsL or I created the respective chromophores. The three atoms of the tetradentate ligands (P_3 or As_3) coordinated in equatorial positions and remaining fourth P, As or N occupied with the unidentate ligands an axial positions. The tetradentate ligands created with Pt(II) atom a five-membered metallocycles with the mean P-Pt-P and N-Pt-P "bite" angles of 85.2 and 84.1°, respectively. The mean Pt-P_{eq} vs Pt-P_{ax} bond distances are 2.350 vs 2.284 Å.

2.3. Pt(III) Complexes

There are fifteen black Pt(III) complexes in fourteen of them a Pt(III) atoms are in a square-planar (PtN_4 (x1), PtS_4 (x12), $PtSe_4$ (x1) and in remaining one in a square-pyramid ($PtCl_3N_2$) environments. Each a square-planar about a Pt(III) atom is created by a pair of homo-bidentate ligands: diphenylglyoximate-N,N; $S_2C_2(CN)_2$-S,S□; $S_2C_2(CF_3)_2$-S,S□; $S_2C_2Ph_2$-S,S□; 4,5-dimercapto-1,3-dithio-2-thionate-S,S□; 5,6-dihydro-1,4-dithio-2,3-dithiolate-S,S□; 5,7-dihydro-1,4,7-trithein-2,3-dithiolate-S,S□or $Se_2C_2(CF_3)_2$-Se,Se□ There homo-bidentates form with a Pt(III) atom a five-membered metallocycles, with the mean L-Pt-L "bite" angles, open in the order: 77° (NC_2N) < 89.5° (SC_2S) < 90.3° (SeC_2Se). The mean Pt-L bond distance elongated with the covalent radius of coordinated atom in the sequence: 1.95 Å (N, 0.75 Å) < 2.275 Å (S, 1.02 Å) < 2.367 Å (Se, 1.16 Å).

A square-pyramid about a Pt(III) atom is created by three Cl and a homo-bidentate ethylenediamine-N,N′ ($PtCl_3N_2$). The equatorial plane about the Pt atom is made up of two N and two Cl atoms. The third Cl atom is apical with a bond length of 2.376(3) Å which is 0.063 Å shorter than the Pt-Cl(eq) at 2.313(3) Å.

2.3. Pt(IV) Complexes

There are over two hundred and twenty monomeric Pt(IV) coordination complexes for which structural parameters are available. Only in three of them Pt(IV) atoms have a square-planar arrangement (PtN_4 (x2) and PtS_4). In all remainders arc hexacoordinated with the following inner coordination spheres: PtX_6 (X=O, N, Cl, S, P, Br, Se or I); PtF_5Sl, PtN_5Cl, $PtCl_5S$ (X=O, N, S or Br), $PtBr_5P$; PtF_4Cl_2, PtO_4X_2 (X=N, Cl, S, Br or I), PtN_4X_2 (X=F, O, Cl, Br or I), PtC_4N_2, $PtCl_4X_2$ (X=O, N, S, P, Se, As or I), $PtBr_4X_2$ (X=N or S), $PtSe_4Cl_2$, $PtSi_4Cl_2$, $PtAs_4X_2$ (X=Cl or I), PtI_4N_2; PtF_3Cl_3, PtN_3X_3 (X=Cl or I), $PtCl_3S_3$; PtN_4ClBr, PtN_4CI, $PtCl_4ON$, $PtCl_4NS$, $PtBr_4ON$, PtO_3N_2Cl, PtN_3O_2Cl, PtN_3Cl_2S, $PtBr_3N_2S$, $PtSi_3P_2H$, $PtO_2N_2Cl_2$, $PtN_2Cl_2S_2$, $PtCl_2P_2Br_2$, $PtCl_3ONS$, $PtCl_2S_2NP$ and PtP_2Br_2NS. These complexes crystallized in the following crystal classes: monoclinic (x97) > orthorhombic (x59) > triclinic (x56) > cubic (x4) > tetragonal (x3) > rhombohedral (x2). There are three Pt(IV) complexes in which the metal atom is four-coordinated. In one of them four unidentate –N ligands create a square-planar geometry (PtN_4). The same chromophore appears in another complex using a terdentate- N,N′,N″ {N-(2-aminoethylidene)-1,2-diaminoethane} ligand with a unidentate –N(1-methylcytosine). A black derivative with the PtS_4 chromophore has a pair of chelating (S,S′) 5,6-dihydro-1,4-dithio-2,3-dithiolate(-2) ligands. The mean Pt-S bond distance is 2.242(7) Å and the mean S-Pt-S “bite” angle is 87.7(3)°.

There are over sixty Pt(IV) complexes where each Pt(IV) atom is six-coordinated by equivalent donor atoms (PtX_6). Two complexes contain a PtO_6, one has a centrosymmetric trans-$[Pt(\eta^2\text{-oxalate})_2(OH)_2]^{2-}$ anion connected via hydrogen bonds from its OH groups to the aqua ligands of two protonated complex anions trans-$[Pt(\eta^2\text{-oxalate})_2(OH)(H_2O)]^-$. While the divalent anion has six almost equal Pt-O bond lengths / four at 1.993(3) Å and two at 1.994(3) Å, the monovalent anion has four Pt-O distances of 1.992(3) Å and the other two are 1.958(4) and 2.034(4) Å. In yellow derivative a pentadentate oxygen donor ligand $\{B_7O_{11}(OH)_5\}$ plus a hydroxide group form a distorted octahedral geometry about the Pt(IV) atom (PtO_6). The pentadentate forms a four six-membered metallocycles with the mean O-Pt-O “bite” angle of 91.5(1)°.

Six yellow Pt(IV) complexes have the PtN_6 chromophore. In two of them a pair of bidentate chelating ligands (1-methylcytosinate)(mecy) or 2,4-pentanediiminate)(pdi) plus two NH_3 molecules form the inner coordination sphere. The mecy ligands form two four-membered metallocycles with the mean N-Pt-N “bite” angle of 63.9(4)°, in another complex the pdi ligands form a six-membered metallocycles (90.3(2)°). In both complexes a pair of chelating ligands create the plane while the NH_3 molecules occupy the axial sites, to give a somewhat elongated tetragonal bipyramidal geometry about the Pt(IV) atoms (4+2). In another yellow complex a terdentate (N,N′,N″) ligand N^1,N^3-dinitrosodiethylenetriamine(-2)., with two unidentate NO_2^- anions and a pyridine created a somewhat elongated tetragonal bipyramidal geometry (4+2). The NO_2^- anions occupy axial positions, and the terdentate ligand with pyridine occupy plane. The terdentate ligand forms two five-membered metallocycles (82.7(4)°). In the remaining PtN_6 species a hexadentate ligand is involved. The ligand make both five- and six-membered metallocycles with the mean N-Pt-N "bite" angles of 84.5 and 92.2°, respectively. The mean Pt-N bond length elongated in the sequence: 2.02 Å (bi-) < 2.035 Å (ter-) < 2.07 Å (uni-) < 2.075 Å (hexadentate). The mean N-Pt-N "bite" angle open with increasing number of atoms as expected: 64° (NCN) < 84° (NC_2N) < 92° (NC_3N).

There are forty examples with the $PtCl_6$ chromophore, mostly yellow or orange with the mean Pt-Cl (equatorial) vs Pt-Cl (axial) bond distances of 2.31 vs 2.32 Å. Six orange-red derivatives contain a PtS_6 inner coordination sphere. In three of them six thiocyanate groups form a pseudo-octahedral arrangement about Pt(IV), with Pt-S bond distance range from 2.355(2) to 2.396(2) Å (average 2.375 Å). Two complexes contain the $[Pt(\eta^2\text{-}S_5)_3]^{2-}$ anion. The S_5^{2-} ligands are coordinated to Pt(IV) in a bidentate 1,5-fashion, forming three six-membered metallocycles around it, with the mean S-Pt-S "bite" angle of 91.2° (range 90.5 – 92.1°). In another derivative three S_6^{2-} ligands in 1,6-fashion forming three seven-membered metallocycles with the mean S-Pt-S "bite" angle of 99.9(2)°. The mean Pt-S bond distance of 2.373(7)° is about 0.021 Å shorter than in the previous example (S_5^{2-}).

Two terdentate cyclic-P_3 and triphos ligands created about Pt(IV) atom a pseudo-octahedral geometry (PtP_6). The mean Pt-P bond distances are 2.435(3) Å (P_3) and 2.339(2) Å (triphos). There are two types of metallocycles, one three and one six-membered (each of pair), with the mean P-Pt-P "bite" angles of 52.6(1) and 90.0(1)°, respectively. Six bromine atoms created a pseudo-octahedral geometry (Pt-Br, range 2.329(2) – 2.440(2) Å). Another three derivatives contain $[Pt(\eta^2\text{-}Se_4)_3]^{2-}$ complex anion. Each bidentate Se_4^{2-} anion coordinated in a bidentate 1,4-fashion and forms a five-membered metallocycle with the mean Se-Pt-Se "bite" angle of 99.4(1)° and the mean Pt-Se bond length of 2.486(2) Å. In another derivative a six iodine atoms created pseudo-octahedral arrangement about Pt(IV), with the mean Pt-I bond distance of 2.666(1) Å, which is about 0.210 and 0.852 Å longer than those for $PtBr_6$ and $PtCl_6$.

There are seventeen Pt(IV) complexes of the composition PtA_5B. In the twelve of them a six-unidentate ligands created a pseudo-octahedral environment about each Pt(IV) atom, namely PtF_5Cl, $PtCl_5X$ (X=O, N, S or P) and $PtBr_5P$. The mean Pt-F (equatorial, trans to F) vs Pt-F (axial, trans to Cl) bond lengths are 1.93 vs 1.96 Å. The mean Pt-Cl bond distance elongated in the order: 2.258 Å (trans to F) < 2.28 Å (trans to O) < 2.303 Å (trans to N) < 2.315 Å (trans to Cl) < 2.340 Å (trans to S) < 2.425 Å (trans to P). The mean Pt-Br bond lengths are 2.468 Å (trans to Br) and 2.553 Å (trans to P). The mean Pt-L (axial) bond distance elongated in the order: 2.04 Å (L=N, trans to Cl) < 2.07 Å (O, trans to Cl) < 2.308 Å (P, trans to Cl) < 2.335 Å (S, trans to Cl) < 2.364 Å (P, trans to Br).

Two yellow Pt(IV) derivatives have composition of PtN_5O. In one of them three unidentate, two NH_3 and 1-methylcytosine (mec) with a bidentate η^2-mec donate N atoms with a hydroxide group form a pseudo-octahedral geometry about the Pt(IV) atom. The chelating ligand forms a four-membered metallocycle with the N-P-N "bite" angle of 65.0(6)°. The other derivative has ter- (tach) and bi- (α,α-bipyridyl)dentate N donor ligands which together with H_2O form a pseudo-octahedral geometry (PtN_5O).

There are four yellow Pt(IV) derivatives with PtN_5Cl chromophore. In two of them three unidentate (two NH_3 with NO_2 group in one, and two pyridine with NO_2 group) ligands with bidentate $N(O)N(CH_2)_2NH_2$ and α,α-bipyridyl with NO_2 group and Cl forms a such chromophore. In the fourth one terdentate (diethylenetriamine), NH_3 and NO_2 with Cl created such inner coordination sphere. The mean Pt-L bond distance in PtN_5X (X=O or Cl) elongated in the order: 2.02 Å (OL) < 2.04 Å (bi-NL) < 2.05 Å (ter-NL) < 2.09 Å (NL) < 2.34 Å (Cl).

There are almost one hundred Pt(IV) derivatives of the general composition PtA_4B_2 with twenty seven variety of the coordinated atoms/ligands. The inner coordination sphere of almost 1/3 derivatives is built up only by unidentate ligands. The cis- and trans-PtF_4Cl_2, trans-PtN_4F_2 (N=pyridine); cis- and trans-PtN_4O_2 (N=2-methylcytosine, OH); trans-PtN_4X_2 (N=NH_3 and X=Cl or Br), trans-PtC_4N_2 (C=CN, NH_3), $PtCl_4O_2$ (O=H_2O, L-alanine); cis-$PtCl_4N_2$ {N=$(NH_3)_2$, (1-hydroxyethylamine)$_2$, (1-ethylimidazole)$_2$, (1-propylimidazole)$_2$}, trans- $PtCl_4N_2$ {N=$(NH_3)_2$, (HN=C(Me)ON=CMe_2)$_2$, (HN=C(Me)ON=C(C_5H_{10}))$_2$, (HN= C(Me) CONEt_2)$_2$; (HN=C(CMe)ON=C(Cl)(C_6H_4-p-Me))$_2$ or (HN=C(CMe)ON=C(Cl) +(C_6H_4-p-NO_2)$_2$}; cis-$PtCl_4S_2$ (S=$(SCN)_2$), trans-$PtCl_4S_2$ (S=$(Me_2S)_2$ or (di(2-ethylhexyl)sulphido)$_2$; cis-$PtCl_4P_2$ (P=$(PEt_3)_2$) and trans-$PtBr_4S_2$ (S=$(Me_2S)_2$).

There are over twenty Pt(IV) derivatives in which a pseudo-octahedral arrangement about Pt(IV) atom is created by one bidentate and four unidentate ligands (PtA_4B_2). Such combination forms the following inner coordination spheres about Pt(IV) atoms: cis-PtO_4N_2 (4OH or, 4(acetate)+(η^2-1,2-diaminocyclohexane-N,N) (dach); cis-$PtCl_4N_2$ (η^2-dach); trans-PtN_4Cl_2 (η^2-1,2-propanediamine-N,N′) (1,2-pn)+NH_3+NO_2; cis-PtN_4Cl_2 (η^2-ethylenediamine-N,N′) (en)+2NO_2; (η^2-N(O)N(CH_2)$_2$$NH_2$)+$NH_3$+$NO_2$; ($\eta^2$-N(O)N($CH_2$)$NH_2$)+$NH_3$+NO; cis-$PtCl_4N_2$ (η^2-α,α′-bipyridine-N,N′(bpy)); η^2-aminomethylpyrrolidine-N,N′) or (η^2-2,2-Me_2-1,3 -propanediamine-N,N′); cis-$PtCl_4X_2$ (η^2-$C_5H_4F_3S_2$-S,S′); (η^2-1,2-(Me_2Se)$_2$$C_6H_4$-Se,Se′); cis-$PtBr_4N_2$ (η^2-en-N,N′); (η^2-1,3-pn-N,N′); trans-$PtSi_4C_2$ (η^2-bsit-Si,Si′)+2SiMe_2Ph+ 2CNBu^t); cis-PtI_4N_2} η^2-1,10-phenanthroline-N,N′(phen)), (η^2-4,7-Ph_2phen) and (η^2-2,9-Me_2-4,7-Ph_2phen).

In another over twenty Pt(IV) derivatives a pseudo-octahedral geometry about Pt(IV) atoms is created by two bidentate ligands with two unidentate with the following mostly with trans-configuration. Trans-PtO_4N_2: (η^2-4-hydroxy-L-prolinate-O,N)$_2$+2OH; (η^2-malonate-O, O′)(mal)+(η^2-pn-N,N′)+2OH; (η^2-mal-O,O′)+(η^2-en-N,N′)+2OH; (η^2-1,1-cyclobutane dicar-boxylate-O,O′) (chbd)+(η^2-en-N,N′)+2OH; (η^2-1,1′-Me_2mal-O,O′)+(η^2-1,3-pn-N,N′)+2OH; (η^2-chbd-O,O′)+(η^2-1,3-pn-N,N′)+2OH; trans-PtO_4Cl_2: (η^2-oxalate-O,O′(OX))$_2$; (η^2-acetyl-acetonate-O,O′)$_2$ (acac))$_2$; trans-PtO_4S_2: (η^2-acac-O,O′)$_2$+2SCN; trans-PtO_4X_2: (η^2-ox-O,O′)$_2$ +Br_2; or I_2; (η^2-acac-O,O′)$_2$+Br_2, or I_2; cis- and trans-PtN_4Cl_2: (η^2-en-N,N′)$_2$; trans-PtN_4Cl_2: (η^2-1,3-pn-N,N′)$_2$; trans-PtN_4X_2: (η^2-en-N,N′)$_2$+Br_2 or I_2; trans-PtN_4I_2: (η^2-1,2-benzoquinone-dioxi-mate-N,N′)$_2$; cis-PtS_4I_2: (η^2-Bu_2dithiocarbamate-S,S′)$_2$; trans-$PtAs_4X_2$: (η^2-1,2-(Me_2As)$_2$. C_6H_4-As,As′)$_2$+Cl_2 or I_2.

There is derivative in which terdentate η^3-diethylenetriamine-N,N′,N″(dien) + unidentate 9-methylquanine-N,+2OH forms trans PtN_4O_2 chromophore. In another cis-PtS_4As_2 two terdentate η^3-As_3S_5 ligands are involved. In six cis-PtN_4Cl_2 (x3) and trans-PtN_4Cl_2 (x3) a tetradentate (η^4-tetraaza-derivatives-N,N′,N″,N‴) ligands with two chlorine atoms form such inner coordination sphere.

The Pt(IV) complexes of the PtA_4B_2 type consist of the twenty seven chromophores: PtF_4Cl_2; PtO_4X_2 (X=N, Cl, S, Br or I); $PtCl_4N_2$; PtC_4N_2; $PtCl_4X_2$ (X=O, N, S, P , or Se); PtS_4X_2 (X=As or I); $PtBr_4X_2$ (X=N or S); $PtSe_4Cl_2$; $PtSi_4X_2$ (X=C or P), $PtAs_4X_2$ (X=Cl or I) and PtI_4N_2. There is wide variety of uni- F, O_4, NL, CN, Cl, SL, PL, Br, SeL, SiL, AsL and I; bi- O_2L, N_2L, S_2L, Se_2L, Si_2L, As_2L; O+NL, ter-N_3L, S_2+AsL and tetra-N_4L , donor sites. The mean Pt-L bond distance (homo-trans) elongated in the order: 1.94 Å (F) < 1.98 Å (CN) < 2.00 Å (O) < 2.05 Å (N) < 2.31 Å (Cl) < 2.355 Å (S) < 2.43 Å (Si) < 2.45 Å (As) < 2.46 Å

(Br) < 2.50 Å (Se) < 2.67 Å (I). The mean Pt-L bond distances (homo trans vs hetero trans) are: L=F. 1.94 vs 1.96 Å (Cl, 2.263 Å); O, 2.00 vs 2.01 Å (N, 2.02 Å) < 2.07 Å (Cl, 2.265 Å); N, 2.05 vs 2.045 Å (Br, 2.450 Å) < 2.055 (Cl, 2.310 Å) < 2.105 Å (I, 2.610 Å); Cl, 2.31 vs 2.34 Å (Se, 2.435 Å) < 2.36 Å (S, 2.338 Å) < 2.695 Å (P, 2.335 Å); S, 2.355 vs 2.34 Å (I, 2.66 Å) < 2.43 Å (As, 2.455 Å); Si, 2.43 vs 2.38 Å (P, 2.408 Å).

The chelating ligands form a variety of metallocycles, and the effect of both steric and hindrance factors can be seen in the opening of the LPt-L bond angles of the metallocycles. The five-membered metallocycles open in the sequence: 79.5° (NC_2N, unsaturated) < 83.8° (OC_2O) < 84.0° (NC_2N, saturated) < 84.5° (OC_2N) < 87.6° (SeC_2Se) < 89.5° (SC_2S) < 99.5° (SAs_2S), for the six-membered metallocycles: 91.1° (SC_3S) < 94.0° (OC_3O) < 94.4° (NC_3N).

There are ten Pt(IV) complexes with the PtA_3B_3 composition, PtF_3Cl_3 (x2), PtN_3Cl_3 (x5), PtN_3I_3 (x1) and $PtCl_3S_3$ (x2), for which crystallographic and structural data are available. Two pairs fac- and mer-PtF_3Cl_3 and fac- and mer-$PtCl_3S_3$ have inner coordination sphere build up by six unidentate ligands (S donor are SCN groups). In other three mer-PtN_3Cl_3 also unidentate ligands, N-donor ligands are two NH_3 molecule and uracilate derivative. In another one mer-PtN_3Cl_3 N-donor ligands are bidentate η^2-en-N,N′ and imidazolidinethiolate. In fac-PtN_3Cl_3, N-donor ligand is terdentate η^3-dien. In fac-PtN_3&I_3, N-donor ligand is terdentate η^3-4-Medien.

Another seven Pt(IV) derivatives have a PtA_4BC composition, PtN_4CX (X=Br or I), $PtCl_4ON$ (x2), $PtCl_4NS$ (x2) and $PtBr_4ON$. Two yellow PtN_4CX (X=Br or I) are isostructural, a pair of bidentate η^2-en-N,N″ ligands, with unidentate CN group and X, created such composition. Two $PtCl_4ON$ which are also isostructural, besides four chlorine atom a bidentate -O,N glycinate completed a pseudo-octahedral arrangement. In another two monoclinic and orthorhombic $PtCl_4NS$, donation N and S ligands are differ, while in monoclinic is NH_3 and Me_4thiourea-S; in orthorhombic is heterobidentate –N,S, 1-methylcysteinsulphoxide. Another heterobidentate –O,N $pyPh_2PO$ ligand with four Br atoms created $PtBr_4ON$ chromophore. Six monoclinic Pt(IV) complexes of PtA_3B_2C composition, PtO_3N_2Cl, PtN_3O_2Cl, PtN_3Cl_2S, $PtBr_3NS$ (x2) and $PtSi_3P_2H$ created with a variety of donor ligands. The PtO_3N_2Cl is created by two $OSiMe_3$, one OH, bidentate –N,N′, 2,2-Me_2propenediamine molecule and Cl atom. One hetero-terdentate –O,O′,N (M-Me iminodiacetate), with one bidentate –N,N′(dach) and a Cl atom created PtN_3O_2Cl chromophore. The PtN_3Cl_2S is from by terdentate –N,N′,N″(dien), unidentate Me_4thiourea and two Cl atoms. Two $PtBr_3N_3S$ build up by heteroterdentate –N,N′,S (tdac, or S_4N_4) and three Br atoms. Finally, the $PtSi_3P_2H$ is forms by three unidentate SiH_2Ph ligands, bidentate –P,P′($Me_2PCH_2CH_2PMe_2$) and a hydride.

An analyzed of PtA_3B_3, PtA_4BC and PtA_3B_2C shows that these types of complexes crystallized in monoclinic (x12), orthorhombic (x5) and triclinic (x4) crystal classes. There are uni- H, F, OL, NL, CN, Cl, SL, Br, SiL and I; bi- N_2L and P_2L, O+NL, and N+SL; ter- N_3L, O_2+NL and N_2+SL donor ligands. The mean Pt-L bond distances (homo-trans vs hetero-trans) are: F, 1.935 vs 1.975 Å (Cl, 2.7 Å); O 1.945 vs 2.202 Å (Cl, 2.30 Å) < 2.045 Å (N, 2.02 Å) < 2.095 Å (Br, 2.405 Å); N, 2.07 vs 2.04 Å (Cl, 2.305 Å) < 2.07 Å (Br, 2.47 Å) < 2.10 Å (I, 2.625 Å); Cl 2.31 vs 2.29 Å (Br, 2.49 Å) < 2.355 Å (I, 2.35 Å); Si, 2.375 vs 2.375 Å (P, 2.374 Å) < 2.405 Å (H, not given). Chelating ligands created five- and six-membered metallocycles, with the mean L-Pt-L „bite“ angles, which open in the sequences: five- 83.3° (NC_2N) < 84.3° (OC_2N) < 85.0° (PC_2P) < 90° (NSNS); and six- 90° (OC_3N) < 91.5° (NC_3N) < 99° (SNSNS).

There are twenty $PtA_2B_2C_2$ type complexes, $PtO_2N_2Cl_2$ (x17), $PtO_2N_2I_2$, $PtN_2Cl_2S_2$ and $PtCl_2P_2Br_2$, which crystallized in orthorhombic (x9), monoclinic (x8) and triclinic (x3) crystal classes. In six complexes with a $PtO_2N_2Cl_2$ and one $PtCl_2P_2Br_2$ chromophores, are created by a unidentate ligands. In another four derivatives with $PtO_2N_2Cl_2$ and one $PtO_2N_2I_2$, N-donor ligands are bidentate, in two examples is η^2-en-N,N′ and in another two is η^2-dach-N,N′ with pairs of unidentate O donor (acetate) and two chlorine or iodine atoms. In remaining $PtO_2N_2Cl_2$ a pair of hetero-bidentate –O,N with two chlorine atoms build up a pseudo-octahedral geometry. A pair of heterobidentate –N,S imidazolidine-thiazole ligands with two chlorine atoms form $PtO_2Cl_2S_2$ chromophore.

There are even four Pt(IV) derivatives in which an inner coordination sphere build up by four different donor atoms, $PtCl_3ONS$, $PtCl_2S_2NP$ and PtP_2Br_2NS (x2). Trans- $PtCl_3ONS$ is created by three chlorine atoms, dmso-S and heterobidentate –O,N (salicylaldoximate) ligand. A terdentate- S,S′,N (η^3-S_4N_4) with PMe_2Ph and two chlorine atoms form $PtCl_2S_2NP$ chromophore. Two cis- PtP_2Br_2NS are created by bidentate –N,S (N(H)SNS) ligand, two PMe_2Ph molecules and two bromine atoms.

Analyzed of $PtA_2B_2C_2$, $PtCl_3ONS$, $PtCl_2S_2NP$ and PtP_2Br_2NS shows that there are uni- OL, NL, Cl, PL, Br and I; bi- O_2L, N_2L, O+NL, N+SL and ter- S_2+NL, dentate ligands, which created such chromophores about the Pt(IV) atoms. The mean Pt-L bond distances are: Pt-O, 2.02 Å (trans O); Pt-N 2.09 Å (trans P); Pt-Cl, 2.305 Å (trans O) < 2.312 Å (trans N) < 2.315 Å (trans Cl); Pt-S, 2.327 Å (trans Br) < 2.337 Å (trans N) < 2.394 Å (trans P); Pt-P, 2.327 Å (trans Br) < 2.346 Å (trans P); Pt-Br, 2.396 Å (trans S) < 2.556 Å (trans P); and Pt-I, 2.327 Å (trans I). Chelating ligands created a variety of metallocycles with L-Pt-L "bite" angles open in the order: 83.5° (NC_2N) < 83.6° (OC_2N) < 84.8° (NC_2S) < 90° (SC_3S) < (OC_4N) < 97° (OC_3O). (NC_3N) < 98.7° (NC_4N).

2.4. Mixed-Valence Pt Complexes

There are three black derivatives, containing two different well separated complex ions, the $[Pt(CNMe)_4]^{+2}$ cation and two $[Pt(\eta^2\text{-mnt})_2]^-$ anions. The complex cation has a Pt(II) atom coordinated to four MeNC ligands with the mean Pt-C bond distance of 1.978(5) Å in a square planar arrangement. The complex anions have a Pt(III) atom with a pair of homobidentate maleonitriledithiolate(-2) (PtS_4) with a mean Pt-S bond distance of 2.265(3) Å and S-Pt-S "bite" angles of 90.0(7)°.

There are over sixty derivatives containing well separated complex ions of a square-planar Pt(II) and a six-coordinated Pt(IV) atoms. In eight derivatives only unidentate ligands created the following inner coordination spheres: Pt(II)N_4 and Pt(IV)Cl_6 (3 derivatives), Pt(II)N_4 and Pt(IV)$(SCN)_6$ (1 derivative), Pt(II)N_4 and Pt(IV)N_4Cl_2 (3 derivatives), Pt(II)N_4 and Pt(IV)N_4Br_2 (1 derivative). The unidentate N-donor ligands are NH_3 and RNH_2. In over forty derivatives a pair of homobidentate –N,N′ ligands, mostly en, 1,2-pn, 1,3-pn and dach, with the respective halogen atoms, created the following inner coordination spheres: Pt(II)N_4 and Pt(IV)N_4Cl_2, Pt(II)N_4 and Pt(IV)N_4Br_2, Pt(II)N_4 and Pt(IV)N_4I_2, Pt(II)Cl_4 and Pt(IV)N_4Cl_2. In two derivatives a terdentate –N,N′,N″ (η^3-dien, or η^3-Medien) with iodine atoms form Pt(II)N_3I and Pt(IV)N_3I_3 chromophores. In another one Pt(II)N_2C_2 and Pt(IV)C_6 chromophores are build up by homobidentate –N,N′ (η^2-dienH) ligand and CN groups. In five remainders the $Pt(II)N_2X_2$ (X=Cl, Br or I) and $Pt(IV)X_4N_2$ (X=Cl, Br or I) chromophores are build up besides halogene atoms by homobidentate –N,N□(mostly en).

The ligands involved in the series of mixed valence complexes are uni-, bi- and terdentate. The mean Pt(II)-L bond distance elongated in the order: 1.97 Å (CN) < 2.03 Å (NL) < 2.065 Å (ter-NL) < 2.07 Å (bi-NL) < 2.305 Å (Cl) < 2.43 Å (Br trans to N) < 2.567 Å (I trans to N) < 2.635 Å (I). The mean Pt(III)-S bond length is 2.265 Å (bi-SL). The mean Pt(IV)-L bond length elongated in the order: 1.99 Å (CN) < 2.03 Å (NL) < 2.06 Å (bi-NL) < 2.08 Å (ter-NL) < 2.315 Å (Cl) < 2.341 Å (SL) < 2.476 Å (Br trans to N) < 2.63 Å (I trans to N) < 2.727 Å (I). Noticeable, the mean Pt(IV)-L bond distances except NL are somewhat longer than those of Pt(II)-L bond distances, as a reason of higher coordination number six against four.

2.5. Pt(2.87), Pt(3.25) and Pt(3.5) Complexes

Three black derivatives contain Pt atoms in non-integral oxidation states and each crystallized in the triclinic crystal class. In three different complex cations, $[Pt(\eta^2\text{-mnt})_2]^{+1.3}$, $[Pt(\eta^2\text{-mnt})_4]^{+0.75}$ and $[Pt(\eta^2\text{-mnt})_2]^{+0.5}$ a pair of homobidentate –S,S□(maleonitriledithiolate(-2)) created a square-planar geometry about each P atom. The mean Pt-S bond lengths are 2.269 Å (Pt(2.87)), 2.271 Å (Pt(3.25)) and 2.266 Å (Pt(3.5)). The mean S-Pt-S "bite" angles are 90.3°.

3. DIMERIC PT COORDINATION COMPLEXES

3.1. Pt(0) Dimers

There are four colored dimeric Pt(0) complexes. In red $[Pt(\eta^2\text{-}Bu^t_2P(CH_2)_3PBu^t_2)]_2$ each Pt(0) atom is three-coordinate (Y-shape), being bonded to two P atoms of chelating ligand and the other Pt(0) atom at a distance of 2.765(1) Å. The mean Pt-P bond distance is 2.270 Å and chelating ligand creates a six-membered metallocyclic ring with the mean P-Pt-P angle of 102.6(1)°. In another red dimer, three bis(diphenylphosphino)methane ligands bridges two Pt(0) atoms with Pt-Pt separation of 3.022(3) Å. Each Pt(0) atom displays trigonal planar geometry (PtP_3). A similar Pt-Pt separation of 3.025(1) Å is found in another red dimer, but here are two non-equivalent Pt(0) atoms, one 3- and the order 4-coordinate. Both Pt-Pt distances suggest a significant metal-metal interaction. In yellow-brown dimer two centrosymmetrically related $Pt(PBu^t_2Ph)$ fragments are held together by two double bridging CS_2 ligand. Each CS_2 group is π-bonded through a C=S bond to one Pt(0) atom, and σ-bonded to the other via the exo S atom. The result is a six-membered metallocycle (-PtCSPtCS-) with an essentially planar coordination about the Pt(0) atoms.

3.2. Pt(I) Dimers

There are over thirty Pt(I) dimers for which structural parameters are available. These dimers crystallized in monoclinic (x19), triclinic (x13) and orthorhombic (x4) crystal classes. There is a direct Pt(I)-Pt(I) bond in a number of colored (mostly yellow, some orange or red) dimers, ranging from 2.5272(5) to 2.791(1) Å. In five of these, two equivalent PtP_3H units are held together only by this metal bond. In two derivatives there is also one bidentate non-chelating ligand supplying extra links between the metal atoms, 2,2□-bipyridyl with two N donor atoms in one and in the other $Ph_2PCH_2PPh_2$ with two P donor atoms. The most common situation is two non-chelating bidentate ligands bridging the two Pt(I) atoms with O+P, N+P, P+P or S+P donor atoms. In another three cases two single bridging ligands (μ-$PPh_2)_2$, (μ-$PBu^t_2)_2$, are involved. A μ-S or μ-Se atom serves as bridge in two other examples. Some orange dimers have two bis(diphenylphosphinomethyl)phenylphosphine ligands as bridges in one case both are homo-terdentate (P-donor), while in the other one is bi- and the other terdentate, indicating the Pt(I) atoms to be either 4- to 5-coordinate.

An analyzed of the Pt(I) dimers shows that in mostly derivatives both Pt(I) atoms are equivalently coordinated, including the Pt-Pt bond, with the inner coordination spheres: PtX_3Pt□(X=N or P); PtP_2XPt□(X=H, O, N, C, CL or S); PtNClPPt□and PtP_2HCPt□. In other derivatives two non-equivalent Pt(I) atoms are found with chromophores: PtP_3 and PtP_2H; PtP_3 and PtP_2X (X=C or P); PtP_2N and PtP_2Cl; PtP_2Se and PtCPSe; PtP_4 and PtP_2C. The four-coordinated Pt(I) atoms are in a square-planar environment with varying degrees of distortion, including the Pt-Pt bond. The five-coordinated Pt(I) atoms mostly are in a trigonal-bipyramidal environment. As was mentioned all Pt(I) dimers contain a direct Pt-Pt bond, which range from 2.527 to 2.791(1) Å (average 2.645 Å). The mean Pt-X-Pt bridge angles open in the order of X: 68.2° (X=Se) < 68.8° (P) < 69.3° (Si) < 71.8° (S) < 84°(C).

The most common type of ligands has P-donor sites (particularly, $Ph_2PCH_2PPh_2$). There are uni-, bi-(N_2, P_2) and ter-dentate (P_3) ligands. The mean Pt(I)-bond distance for the unidentate ligands elongated in the order: 1.69 Å (L=H) < 1.90 Å (CNR) < 2.125 Å (NL) < 2.285 Å (PL) < 2.405 Å (Cl). For homo bi- and terdentate ligands the order is: 2.075 Å (bi-NL) < 2.27 Å (bi-PL) < 2.355 Å (ter-PL). The mean Pt(I)-L(bridge) bond distance elongated in the order: 2.268 Å (S) < 2.31 Å (PL) < 2.374 Å (Se) < 2.376 Å (SiL). There are two types of heterobidentate ligands with O+P and N+P donor atoms. The L-Pt-L "bite" angle opens in the order: 79° (NC_2N) < 88.8° (PC_2P) < 99.5° (PC_3P) < 103.5° (PC_4P).

3.3. Pt(II) Dimers

There are over two hundred and forty Pt(II) dimers, the structures are complex and can be divided into several groups. There are over fifty dimers in which the Pt-Pt bond distance range from 2.6473(5) to 2.9915(4) Å. Only in two examples black (PtS_4) and yellow (PtH_2P_2) the units are held together only by the Pt-Pt bond at 2.748(2) and 2.768(2) Å, respectively. In all other examples there are also bridging ligands in addition to a significant Pt-Pt interaction. In nine derivatives two single atoms double-bridge the metal atoms forming a four-membered metallocyclic core, (-P-Pt-P-Pt-), (-H-Pt-H-Pt), (-H-Pt-S-Pt-), or (-H-Pt-P-Pt-). In another eight derivatives two Pt atoms are bridged by four-chelating bidentate ligands in a 1,3-fashion with N,N; N,S; S,S or P,P donor sites. By far the most common situation is two non-chelating bidentate ligands bridging the metal atoms in a 1,3-fashion. Some have identical donor atoms, N,N or S,S, while the remainders use different atoms O,N with head - to - tail arrangement, or head - to - head arrangement. In one derivative two terdentate ligands using O,N,S donor sites coordinated to each Pt(II) atom.

The Pt(II)-Pt(II) bond distance range from 2.647 to 2.991 Å (average 2.862 Å) and Pt-X-Pt bridge angle opens in the order of an decreasing covalent radius of the respective X: 72.9° (X=P, 1.06 Å) < 74.5° (S, 1.02 Å) < 81.8° (N, 0.75 Å) < 99.7 (H, 0.32 Å). The equivalent inner coordination sphere about the Pt(II) atoms are: PtN_4, PtS_4, PtN_3O, PtP_3H, PtH_2P_2, PtO_2N_2, PtN_2S_2, PtH_2PSi, PtN_2OS, PtN_2ClS, PtP_2HS and PtONClS. There are some derivatives that contain non-equivalent inner coordination spheres for each Pt(II) atom: PtN_4 and PtO_2N_2; PtN_4 and PtO_2NC. The wide range of ligands includes uni-, and terdentate. For unidentate ligands the mean Pt(II)-L bond distance elongated in the sequence: 1.57 Å (H, 0.37 Å) < 2.05 Å (NL, 0.75 Å) < 2.18 Å (SL, 1.02 Å) < 2.25 Å (PL, 1.06 Å) < 2.305 Å (Cl, 0.99 Å) < 2.335 Å (SiL, 1.17 Å). Except for the S-donor ligands, the sequence follows the increasing covalent radius of the donor atom. For the homo-bidentate ligands the order is: 2.005 Å (NL) < 2.05 Å (OL) < 2.28 Å (PL) < 2.31 Å (SL). The mean Pt-L (bridge) bond distance elongated with increasing donor atom covalent radius in the order: 1.80 Å (H) < 2.08 Å (N) < 2.29 Å (S) < 2.305 Å (P). There are two types of heterobidentate ligands, the more common O plus N, and the other with N plus S, donor sites. The terdentate ligands have 3N, 2N+S or O+N+S donor sites. There are several types of metallocycles and the respective L-Pt-L bite angles open in the sequence: 73.5° (SCS) < 82.0° (NC_2N) < 86.0° (PC_2P) < 88.0° (SC_2S) < 90.0° (NC_3N) < 97.0° (PC_3P). In the cases where two single atoms serve as bridges, the Pt-L-Pt angle opens and the μL-Pt-μL angle closes with decreasing covalent radius of L, with the mean values being: 72° vs 168° (L=P) < 74.5° vs 105.5° (S) < 100.5° vs 74.5° (H). The Pt(II)-Pt(II) distances ranging from 2.647 to 2.991 Å, can be subdivided into five groups from a bridge point of view. The mean Pt-Pt distance elongated in the order: 2.73Å (two single atom bridges) < 2.82 Å (four homo-bidentate non-chelating bridges) < 2.88 Å (two hetero-bidentate non-chelating bridges) < 2.925 Å (four hetero-bidentate non-chelating head-to-head bridges) < 2.934 Å (four hetero-bidentate non-chelating head-to-tail bridges).

In all remainders Pt(II) dimers a Pt(II)-Pt(II) separations are over 3.0 Å. Their structures are very complex and can be divided into several groups from the bridges point of view. There are eighty dimeric Pt(II) coordination compounds having two single atoms linking the metal atoms. In sixteen (yellow and colorless) two OH groups serve as bridges. The two OH groups bring the two metal atoms within the range 3.0496(9) to 3.270(1) Å (ave. 3.155 Å). The Pt-O-Pt bridge angles range from 94° to 105.5° (average 100°). As the Pt-O-Pt angle opens the Pt-Pt distance elongated for some but not all, of the examples. In addition to the two O atom bridges, each Pt(II) atom has two other donor atoms: Cl plus S; two N, two P, two S or two As, creating a square planar arrangement with varying degrees of distortion.

Six colored (red, yellow, green) monoclinic species have two N donor atoms as bridges Pt-$(NL)_2$-Pt, with Pt-Pt separation from 3.087(1) to 3.23 Å (average 3.16 Å). The Pt-N-Pt bridge angles range from 93.8(5) to 110(1)° (average 103.2°), with the mean Pt-N (bridge) bond distance of 2.06 Å. Each Pt(II) atom has a square-planar geometry (PtN_2P_2, PtN_2SP, PtN_2SeP). The mean Pt-L bond distance elongated with the increases covalent radius of coordinated atom, in the order: 2.222 Å (SL, 1.02 Å) < 2.255 Å (PL, 1.06 Å) < 2.334 Å (SeL, 1.16 Å).

In another thirteen Pt(II) dimers a pair of Cl atoms provide the bridging Pt-$(Cl)_2$-Pt. The Pt-Cl-Pt bridge angles range from 94.5(1) to 102.7(1)° (ave. 96.4°). Only in two examples the Pt-Pt separation is given, 3.488(2) and 3.481(2) Å, respectively. Each Pt(II) atom is coordinated by four donor atoms: $PtCl_4$, $PtCl_2N_2$, $PtCl_3S$ (X=N, S, P, As) and $PtCl_2P_2$, in a distorted square planar arrangement. The mean Pt-Cl distances are 2.28 Å (terminal) and 2.33 Å (bridge). Both are somewhat shorter than Pt-Cl (trans to P) with the mean value of 2.425 Å as a result of the trans influence of the P atom. The mean Pt-L bond distance elongated with an increases covalent radius of donor atom in the order: 2.02 Å (NL) < 2.188 Å (SL) <2.214 Å (PL) < 2.222 Å (bi-PL). The metallocycles are: 68.4° (PCP) < 97.9° (PC_3P).

There are five (three dark green and two pale yellow), derivatives which have both a single Cl atom and another donor atom acting as bridges, N, S or Te. The Pt-Cl-Pt bridge angle opens and the Pt-X-Pt angle closes in the order: 82.2(3)° and 119(2)° (X=N), 84.5(2)° and 93.1(2)° (S). 99.8° and 92.93(5)° (Te). The Pt-Pt separations range from 3.287(3) to 3.312(2) Å.

The most common single atom acting as a bridge in this series is the S atom or SL donor ligands. There are twenty derivatives in which two S donor atoms serve as a bridge between two PtL_2 moieties ($Pt(S)_2Pt$). Each Pt(II) atom has an approximate square planar arrangement: PtS_3P, PtS_2X_2 (X=N, P (most common), or Br); PtS_2HP, PtS_2NX (X=P or Br), PtS_2ClP and PtS_2PI. The Pt-Pt separations range from 3.141(1) to 3.627(1) Å (ave. 3.39 Å). The Pt-S-Pt bridge angle range from 87.3(1)° to 102.1(1)° (ave. 94.2°). The mean Pt-S (bridge) bond distance of 2.30 Å is shorter than the Pt-S value of 2.37 Å (trans to P) or 2.45 Å (trans to H), because of the trans influence. The mean Pt-L bond distance elongated in the order: 2.01 Å (bi-NL, unsaturated) < 2.04 Å (NL) < 2.08 Å (bi-NL, saturated) < 2.260 Å (bi-PL) < 2.28 Å (PL) < 2.330 Å (SL, terminal) < 2.430 Å (Br) < 2.620 Å (I). The mean N-Pt-N "bite" angles are 90.5° (NC_2N, unsaturated) and 82.7° (NC_2N, saturated). The value of P-Pt-P "bite" angle is 84.8° (PC_2P).

There are ten pale yellow Pt(II) dimers in which a pair of P atoms form the $Pt(PL)_2Pt$ bridges. Each Pt(II) atom has an approximately square planar environment. The Pt-Pt separation in this group ranges from 3.5454(5) to 3.70 Å (ave. 3.635 Å), and the Pt-P-Pt bridge angle ranges from 101.7(1)° to 105.4(1)° (ave. 103.5°). The two values are related, both increasing together. It is noted that the Pt-P and Pt-P (bridge) bond distances are almost the same, with the values of 2.323 and 2.319 Å, respectively.

In dark brown Pt(II) dimers a pair of bromine atoms ($Pt(Br)_2Pt$) join two $PtBr_2$ moieties and completed a square planar environment for each Pt(II) atom with Pt-Pt separation of 3.552(2) and 3.59 Å, and Pt-Br-Pt bridge angles of 93.8 and 94.6(3)°. The mean Pt-Br (terminal) vs Pt-Br (bridge) bond distances are 2.423 vs 2.437 Å.

There are two cream colored Pt(II) dimers where a pair of single Se atoms or EtSe ligands serve as a bridges between two PtP_2 or PtClP moieties, with mean Pt-Se-Pt bridge angles of 100.39(5) and 96.9(1)°, respectively.

A pair of cis- and trans yellow derivatives, contain a pair of SicyH ligands which link two $Pt(PEt_3)_2$ units, completing an approximate square planar geometry about each Pt(II) atom ($PtSiP_2$). The Pt-Pt separation of 3.661(1) Å and Pt-Si-Pt bridge angle of 99.1(2)° found in the cis form differ from those found in the trans form at 4.046(1) Å and 115.5(2)°.

In seven colored Pt(II) dimers a pair of iodine atoms bridge the two Pt(II) atoms to give a central $Pt(I)_2Pt$ core ring with mean Pt-I-Pt angle of 94./8(1)° and mean μI-Pt-μI angle of 85.2(1)°. The Pt-Pt separation ranges from 3.7956(7) to 3.836(2) Å. A pair of Te atoms join two PtP_2 units to form a central $Pt(Te)_2Pt$ core ring with Pt Te Pt and µTe-Pt-µTe angles of 101.3° and 78.7°, respectively.

In this series of Pt(II) dimers are uni-, bi- and terdentate ligands. The most common ligands are P and S donors. The mean Pt-L bond distance for the unidentate ligands elongates with increasing covalent radius of the donor atom, except in the case of Cl for which 40% of the cases are showing a P-trans effect. The values elongate in the sequence: 2.04 Å (N) < 2.22 Å (S) < 2.265 Å (P) < 2.32 Å (Cl) < 2.425 Å (Br) < 2.59 Å (I). The mean Pt-L (bridge) bond distance elongates in the order: 2.07 Å (OH) < 2.32 Å (PL) < 2.34 Å (SL) < 2.36 Å (Cl) < 2.395 Å (SiL) < 2.435 Å (Br) < 2.445 Å (SeL) < 2.58 Å (I) < 2.61 Å (Te). Here except for Cl, PL and SiL, the order reflects the covalent radius of the donor atom.

The effect of both electronic and steric factors of the coordinating atoms and ligands can be seen in the opening of the L-Pt-L bond angles of the respective metallocycles. The L-Pt-L bite angle opens in the sequence: 71.3° (PCP) < 82.0° (NC_2N) < 84.5° (PC_2P) < 87.5° (NSNS) < 91.3° (NSeNSe) < 93.1° (PC_4P) < 94.5° (PC_3P). The mean Pt-Pt separation for different pairs of X bridging atoms elongated in the order: 3.155 Å (X=O, 100°) < 3.165° (N, 103°) < 3.39 Å (S, 94.2°) < 3.485° (Cl, 96.5°) < 3.57 Å (Br, 94.2°) < 3.63 Å (P, 103.5°) < 3.76 Å (Se, 98.5°) <3.82 Å (I, 94.8°) < 3.85 Å (Si, 107.3°) < 3.965 Å (Te, 99.2°). The value in parentheses is corresponding mean value of the Pt-X-Pt bridge angle.

There are three Pt(II) dimers with a single atom and a two-atom bridge between two Pt(II) atoms. In pale yellow derivative two $Pt(PPh_3)_2$ units are bridged by one hydroxo nad one peroxo group forming a five-membered metallocycle (-PtOOPtO-). The mean Pt-O bond distances are 2.087(1) Å (µ-OH), and 2.001(1) Å (µ-peroxo). The Pt-Pt separation of 3.475 Å is too long to permit a bond. Each Pt(II) atom exhibits a distorted square planar geometry (PtO_2P_2). In black derivative $Pt_2(PhNO)_3(PMe_3)_3$ are three distinct modes of nitroso bonding to platinum. One is serve as a unidentate –N and the other as bidentate –O,N bridges and created a five-membered metallocycle (-PtONPtN-). The third nitroso ligand terminally coordinated to Pt(II) atom. The geometry about the Pt(II) atoms, which have different chromophores of PtN_3P and PtP_2ON is distorted square planar. The Pt-Pt separation is 3.502(1) Å and the Pt-N-Pt bridge angle is 111.4(5)°. In another black derivative two non-equivalent PtP_2 and PtClP moieties are double bridged by a single Cl atom and a bidentate non-chelating $P(O)F_2$ group via O,P donor sites. The central metallocycle is puckered and five-membered (-PtOPPtCl-). Each Pt(II) atom has an approximately square planar arrangement, PtP_3Cl and PtP_2OCl. In yellow Pt(II) dimers one single OH group with acetate –O,O□group in a syn-syn arrangement and form a six-membered puckered metallocycle (-PtOPtOCO-). The Pt-Pt separation is 3.1164(6) Å. In orange derivative a single S atom and a $Ph_2PCH_2PPh_2$ ligand form a six-membered central metallocycle (-PtSPtPCP-) with Pt-Pt separation of 4.108(2) Å and Pt-S-Pt angle 125.4(2)°. A distorted square planar environment about each Pt(II) atom is completed by a chelaging dppm ligand (PtP_3S). The mean P-Pt-P bite angle is 72.1(2)°.

There are eleven Pt(II) dimers in which a pair of two atoms ligands form a central six-membered metllocycle (-PtXXPtXX-). In four of the dimers a pair of pyrazole (or its substituent's) create a six-membered puckered $(-PtNN-)_2$ ring. The Pt-Pt separation ranges from 3.156(3) to 3.483(6) Å, with each Pt(II) adopting a near square planar geometry PtN_4 and PtN_2ClX (X=S or P). In yellow Pt(II) dimer two terdentate $Ph_4P_2N_4S_2$ ligands act a chelating (N,S) to each Pt(II) and via the other S□atom forming a central $Pt_2N_2S_2$ ring adopts a chair conformation with the Pt atoms in the 1,4 positions. There are pairs of five- (PtN_2PS)

and seven-membered (PtN_3PS_2) rings with N-Pt-S and S-Pt-S bite angles of 83.4(2)° and 172.9(1)°, respectively. A terminal PPh_3 ligand about each Pt(II) atom completed a square planar geometry PtS_2NP. In colorless complex, two {Pt(PBu^t_3)(H)} units are linked by bridging $SPPh_2$ anions via S,P donor sites and form central six-membered $(-PtSP)_2$ ring. The Pt-Pt separation is 3.620(1) Å. Each Pt(II) atom has PtS_2PH chromophore. Two white dimers contain a six-membered $(PtOS)_2$ ring in chair conformation, each Pt(II) atom is coordinated by two phosphine ligands, PPh_3 or PMe_2Ph, a S and an O atoms of two SO_3^{-2} groups. The mean Pt-Pt separation is 4.29 Å.

In another colorless dimer two nominally square planar Pt centres are linked by a pair of $P(O)Ph_2$ ligands via PO bridges to form a six-membered $(PtOP)_2$ ring in a boat conformation. In a yellow dimer two $Pt(Pr^i_2NCS)_2$ units are held together by a pair of S-P bridges of two $SPPh_2$ ligands to create six-membered $(PtSP)_2$ ring in a chair conformation. Each Pt(II) atom adopts a distorted square planar geometry (PtS_3P). In another dimer, two equivalent Pt(η^2-dppe) units are bridged by a pair of $Si(H_2)Si(H_2)$ ligands and also form a six-membered Pt_2Si_4 ring in a chair conformation. Each Pt(II) atom adopts a square planar geometry (PtP_2Si_2).

There are over twenty Pt(II) dimers in which two non-chelating bidentate ligands serve as a bridges and form a eight membered central rings of the compositions $(PtOCO)_2$, $(PtOCN)_2$, $(PtNCN)_2$, $(PtNCS)_2$, $(PtNCP)_2$, $(PtPCP)_2$, and $(PtPPP)_2$. The Pt-Pt separation range from 3.008(2) to 4.362(2) Å (ave. 3.27 Å). Each Pt(II) atom has a distorted square planar geometry (PtN_4, PtN_3O, PtP_3Cl, PtN_2X_2 (X=O, S or P), PtP_2X_2 (X=C or Cl), PtO_2ClP, PtP_2NS, PtONClS and PtNClSP). There is an example in which the 18-membered ring macrocycle acts as a binuclearing agent, coordinating each Pt(II) centre via one asa and two thia donors, with two $SCH_2CH_2NCH_2CH_2S$ moieties bound meridionally to the Pt(II) centres. The square planar geometry is completed by a terminal Cl atom (PtS_2NCl). The ten-membered central ring $(PtSC_2S)_2$ has a Pt-Pt separation of 4.2807(6) Å. In another two Pt(II) dimers, two PtX_2 (X=Cl or I) are connected by a pair of 1,2-bis(diphenylphosphino)acetylene and form a ten-membered central $(PtPC_2P)_2$ ring. In another dimer, two $PtCl_2$ units are double bridged by a pair of bidentate meta-bis(diphenylphosphino)benzene ligands, which create a twelve-membered ring $(PtPC_3P)_2$ with a Pt-Pt separation of 4.840(8) Å. Two PtX_2 (X=Cl or I) are connected via double bridges of a pair of 2,6-bis(diphenylphosphino)pyridine ligands, also giving a twelve-membered metallocycle $(PtPCNCP)_2$ ring with Pt-Pt separation of 8.20 Å. A pair of resorcinalbis(cyclophosphites) double bridged two $PtCl_2$ units and form a sixteen-membered metallocycle. A sixteen-membered metallocycle also forms with a pair of glutarates and two $Pt(NH_3)_2$ units. Another sixteen-membered rings are formed by a pair of 4,4□dipyrazolylmethane and two $PtCl_2$, as well as two $Pt(PPh_3)_2$ units and a pair of isophthalate. In all of the dimers, each Pt(II) atom has a distorted square planar geometry: PtX_2P_2 (X=Cl or I), PtO_2N_2, PtN_2Cl_2 and PtO_2P_2. The Pt-Pt separations are 8.882 and 8.90 Å.

In yellow Pt(II) dimer two very large $C_{190}H_{116}F_{12}N_4O_{52}S_4$ ligands with two $Pt(PEt_3)_2$ units form a makro-membered ring, with Pt-Pt separation of 10.26 Å. Each Pt(II) has a square planar geometry (PtN_2P_2) but with significant deviation from the ideal 90° angles. The N-Pt-N angles are 81.6(3)° and P-Pt-P angles are 97.2(2)°.

Two Pt(η^2-en) units are double-bridged by a pair of 5-cytidine monophosphate ligands giving a twenty-membered metllocycle. The Pt(II) atoms are slightly distorted square planar (PtN_3O). Another twenty-membered metallocycle is formed by a pair of calix(4)-$(PMe_2)_4$ ligands with two $PtCl_2$ units. The Pt-Pt separation is 7.324(2) Å. The Pt(II) atoms are again in a slightly distorted square planar environment ($PtCl_2P_2$). A twenty-two atoms boats in another colorless Pt(II) dimer which is the largest metallocycle. Here, two Pt(η^2-en) units are double bridged by a pair of 1,4-bis(hypoxanthin-9-yl)butane ligands. Four N atoms about each Pt(II) create a slightly distorted square planar geometry.

There are six Pt(II) dimers in which Pt atoms are triple bridged, two PtH units by three non-chelating bidentate $Et_2PCH_2PEt_2$; in another dimer two $(PhCH_2S)Pt$ units by one S atom and two non-chelating bidentate $Ph_2PCH_2PPh_2$; two PtCl units by one NN-pC_6H_4OMe and two $Ph_2PCH_2PPh_2$; in another dimer two PtCl units by μ-NO and two $Ph_2PCH_2PCH_2$; and two PtCl units by μ-SO and again by two $Ph_2PCH_2PCH_2$ ligands. The Pt-Pt separations range from 3.059(1) to 3.246(3) Å. Each Pt(II) atom has a square planar environment,PtP_3H, PtS_2P_2, PtP_2NCl (in three examples) and PtP_2ClS. The mean Pt-X-Pt (X= NO or S) bond angles are 112.8° and 88.7°.

Two Pt(II) atoms in another two dimers are quadruple bridged by four oxalate ligands or by four pyrazolate ligands. In the former (PtO_4) the Pt-Pt separation of 3.061(2) Å is somewhat shorter than in the latter (PtN_4) with the value of 3.370(1)°.

In the series of Pt(II) dimers beside a double, triple and quadruple bridged are also a single bridged with a variety of atoms: Pt-X-Pt, Pt-XX-Pt, Pt-XXX-Pt and Pt-XXXX-Pt, respectively. There are seven examples of Pt-X-Pt type. Two $Pt(H)(PEt_3)_2$ units are held together by H atom. In another three examples S donor atom of the respective ligand serve as a bridge in one example between two $PtCl_3$ units, in another between $PtCl_3$ and $PtCl_2$ and in the third one between two $PtCl(PEt_3)$ units. In another derivative two non-equivalent $PtCl_2$ and PtP_2Cl moieties are bridged by a $Pr^iPCH_2PPr^i{}_2$ ligand via a P atom. The other P atom of the ligand is attached to the $PtCl_2$ unit. In another two derivatives a single iodine atom serve as bridge between two [η^4[16]aneS_4]Pt units in one derivative and between two $Pt(H)(PPr^i{}_3)_2$ units in another one. Only for two derivative the Pt-Pt separation are available, 3.028(1) Å (μ-H) and 3.985(1) Å (μ-I). The mean Pt-X-Pt bridge angle opens in the order of an increasing covalent radius of X: 114° (X=S, 1.02 Å) < 120° (P, 1.06 Å) < 175° (I, 1.33 Å).

There are four Pt(II) dimers of the Pt-XX-Pt type bridge. In three of them a non-chelating bidentate –N,N□ (pyrazole, adamantane-di-aziridine, hexane-diazilidene) ligand serve as bridge between two Pt (terpyridine) units in the former and between two $PtCl_2(PEt_3)$ units in two remaining. In the four one derivative, two non-equivalent $PtCl_2(PMe_2Ph)$ and $PtCl_2$ units are connected by Se_2Bz_2 ligand in the manner Pt-SeSe-Pt.

In eight colored Pt(II) dimers two Pt atoms are connected by Pt-XCX-Pt type bridge. In six of the dimers non-chelating-NCN□ligands, and in remaining two non-chelating -PCP□ ligands serve bridge between the Pt(II) atoms. The Pt-Pt separation in Pt-NCN-Pt range from 3.035(1) to 3.337(1) Å (ave. 3.613 Å) and in Pt-PCP-Pt the values are 3.747 and 4.671 Å, respectively. There are eleven Pt(II) dimers of the Pt-XXXX-Pt type bridge, ten of them contain Pt-NC_2N-Pt type and remaining one Pt-SC_2S-Pt.

Single bridges Pt(II) dimers crystallized in monoclinic (x17) > triclinic (x7) > orthorhombic (x5) > tetragonal (x1) crystal classes. Each Pt(II) atom in this series has a square-planar environment with the inner coordination spheres: PtN_4 (most common), PtN_3S (X=O, Cl, or S), $PtCl_3S$, PtP_3Cl, PtH_2P_2, $PtCl_2P_2$, $PtCl_2NX$ (X=S or P), $PtCl_2PX$ (X=S or Se), PtS_2ClP and PtP_2HI. In one example the PtS_4I has a square-pyramidal arrangement.

There are twenty Pt(II) dimers which have a unique bridge to each. For example a yellow derivative has a $Pt(PPh_3)_2Cl$ and $Pt(PPh_3)_2$ units held together by a CS_2 ligand which uses all three atoms. The C atom is coordinated to the $Pt(PPh_3)_2Cl$ and both S atoms are coordinated to the $Pt(PPh_3)_2$ unit. In another derivative $\{P(OEt_2O\}_2PO$ ligand serve as bridge between

$PtCl_2$ and $Pt(PEt_3)_2Cl$ units. The ligand coordinates via its two P-donor atoms to the former and by the third P atom to the latter unit. In triclinic two $Pt(NH_3)_2$ units are held together by glycylglycinate as a tetradentate O_2N_2 ligand. One P atom is bonded via terminal O and a peptide N atom and the second Pt is chelated by the peptide O and a terminal N atom. In yellow derivative, two equal $PtCl(PEt_3)$ units are bridged by 2,2-bis(methylthio)ethene-1,1-dithiolate ligand which act as a tetradentate with S-donors sites. Each Pt atom bonded to a chelating ligand, via one end a thiolate S atom and the other via a thioether S atom.

In colorless dimer tetrabis(diphenylphosphino)cyclobutane is the bridge between two $PtCl_2$ units. The ligand acts as a tetradentate chelating each Pt(II) atom by a pair of P-donor atoms. Two $Pt(PPh_3)_2$ unit in yellow dimer are held together by 5-aminoorolate ligand coordinating to one Pt via a carboxylate O atom and an amino N atom and the other Pt by the second carboxylate O atom and amido N atom. In orange dimer, two $PtCl(PEt_3)$ units are held together by 2-(1,3-dioxo-2-indanylidene)benzimidazoline tetradentate ligand, coordinating via O and N atoms giving a six-membered ring on each Pt atom.

In a colorless dimer, a single terephthalate ligand adopts a binuclearing role, bridging two $Pt(PPh_3)_2(NH_3)$ moieties. A bidentate 2,2′-phenylene-bis(4-methyloxazolone) ligand acts as a long bridge between two $PtCl_2(PEt_3)$ moieties. In triclinic derivative two Pt(II) atom are surrounded by two terdentate 6,6′-bis(1-methylbenzimedazol-2-yl)-2,2′-bipyridine ligands and linked by one 1-methylbenzimidazolyl group in the ligand (PtN_4).

A tris-pyrazolylborate ligand bridges $Pt(PEt_3)Br$ and trans-$Pt(PEt_3)Br_2$ units with bidentate coordination to the former and unidentate to the latter. In white derivative two $PtH(PPh_3)_2$ moieties are bridged by bidentate tetracyanodiimidazole ligand. Red derivative involves a tetrathionaphthalene ligand bridging two $Pt(PPh_3)_2$ moieties via cleavage of both S-S bonds. The tetragonal and triclinic dimers have two $PtCl_2$ units bridged by a tetraphos-P_4 ligand in an open mode with a Pt-Pt separation of 6.338(1) Å and 6.916(1) Å, respectively. In a yellow dimer two $PtCl_2(PEt_3)$ moieties are attached to opposite (distal)N atoms of $P_2N_4Se_4$ ring. In the series of dimers with an „unique" structures, are uni-, bi-, ter- and tetradentate ligands, with P and Cl atoms being the most common. Each Pt(II) atom adopts a square planar arrangement with various degree of distortion. The mean unidentate Pt-L bond distance elongates in the order: 2.06 Å (NL) < 2.25 Å (PL) < 2.293 Å (Cl) < 2.355 Å (Cl, trans to P) < 2.42 Å (Br). For polydentate ligands the order is: 2.00 Å (tetra- NL) < 2.01 Å (ter- NL) < 2.065 Å (bi- OL) < 2.11 Å (ter- NL), trans to P) < 2.13 Å (bi- NL, trans to P) < 2.20 Å (ter- PL) < 2.22 Å (tetra- PL) < 2.25 Å (bi- PL), 2.25 Å (tetra- SL) < 2.315 Å (tetra-SL, trans to P). There are hetero-terdentate (C,S,S), hetero-tetradentate (O,N,N,N and O,O,N,N) ligands. The L-Pt-L bite angle opens in the order: 72.5° (SCS) < 80.0° (OC_2N) < 82.0° (NC_2N) < 87.0° (PC_2P) < 88.2° (SC_3S) < 88.8° (SC_2S) < 89.5° (OC_3N) < 97.5° (POPOP).

3.4. Pt(2.5) Dimers

There are two derivatives which contain Pt atoms in a less common oxidation state of +2.5. The yellow derivative has two $Pt(NH_3)_2(NO_2)$ moieties held together by a metal-metal bond (2.584(1) Å. In addition two 1-methylcytosinate ligands bridge as bidentate via the N^3 atom to one Pt and the N^4 atom to the other Pt-atom. The coordination about each Pt(2.5) atom is distorted octahedral (PtN_5Pt). The dark violet derivative has a Pt-Pt bond distance of

2.677(2) Å. The four dithioacetate ligands binding the Pt(2.5) atoms in a syn-syn arrangement creating four three-membered (-SCS-) rings. The dimeric units, $Pt_2(CH_2CS_2)_4$ are also connected by iodine atoms in a linear chain $-Pt_2S_8$-I-Pt_2S_8-I-, staking along the crystallographic b axis. The mean Pt-I bond distance is 2.978(3) Å.

3.5. Pt(III) Dimers

There are almost sixty Pt(III) dimers and all having a direct metal-metal bond. These derivatives can be divided into four structural groups with four bidentate bridges ligands; two bidentate bridges ligands; without bridging ligands; and with a pair of single atom bridges. The four bidentate ligand group with a syn-syn arrangements ($Pt(XCX)_4Pt$) is the most common. There are six types of ligand X donor sites both O atoms (most common), both N atoms; O plus N, N plus S; both S atoms, and both P atoms. The Pt-Pt bond distance in this series range from 2.390(1)° to 2.766(1) Å. The mean Pt-L (bidentate bridge ligand) distance elongates in the order: 2.01 Å (OL) < 2.335 Å (SL) < 2.36 Å (PL). I the heterobidentate the order is: 2.02 and 2.04 Å (O + NL) < 2.09 and 2.30 Å (N + SL). The Pt(III) atoms adopt an unsymmetrical tetragonal bipyramidal geometry, where a tetragonal plane is formed by four donor atoms of four bridging ligands. An axial position is occupied by a unidentate ligand, and the other axial position by the other Pt atom. The mean Pt-L for the unidentate ligand in an axial position elongated in the order: 2.135 Å (OL) < 2.145 Å (NL) < 2.455 Å (Cl) < 2.465 Å (SL) < 2.595 Å (Br) < 2.750 Å (I) corresponding to the covalent radii of the donor atoms.

The two bidentate bridging ligand group is the second most common type, the bridge donor atoms are only O plus N except for one case in which 1-methyl-cytosinate ligands coordinate only via the N atoms. The structures of these derivatives have head-to-tail (HT) and head to-head (HH) linkage of the bridging ligands. The mean Pt-L bond distances found in the HT vs HH examples are 2.017 Å (O) and 2.055 Å (N) vs 2.018 Å (O) and 2.032 Å (N), respectively. The mean Pt-L bond distance for the other ligands around the Pt(III) atoms elongates in the order: 2.055 Å (NL) < 2.0195 Å (OL) < 2.365 Å (Cl) < 2.570 Å (Br). The Pt-Pt bond distances in this group range from 2.527(1) Å to 2.580(1) Å for the HT type and from 2.543(1) Å to 2.650(1) Å for the HH type.

There is one example in which two $PtH(SiMe_2)_2$ units are bridged by two Si atoms of the respective ligands and form a central ($Pt(Si)_2Pt$) ring. The Pt-Pt separation is 2.708(1) Å and the mean Pt-Si-Pt angle 69.5(3)°. There are five Pt(III) dimers in which two independent units are linked only by a metal-metal bond: $PtCl_3N_2$, PtN_4Cl or PtN_4. The Pt-Pt bond distance range from 2.694(1) Å to 3.034(4) Å.

The Pt(III) dimers crystallized in monoclinic (x34) > triclinic (x12) > orthorhombic (x10) crystal classes. The inner coordination spheres about the Pt(III) atoms are: PtN_4Pt and PtN_2S_2Pt□(square-pyramidal) with Pt□atom in apical position. A unsymmetrical tetragonal bipyramids by far prevails with equivalent inner coordination sphere about both Pt(III) atom in the dimers: PtO_5OPt□; PtO_4XPt'' (X=N, S, or Cl); PtN_4XPt□(X=O or Cl); PtS_4IPt□; PtP_4 XPt (X=N, Cl, S , Br or I); PtO_3N_2Pt□, PtN_3O_2Pt□, $PtCl_3N_2Pt$, PtN_3OXPt□(X=Cl or Br); PtS_3NClPt□, PtO_2N_2SPt□(X=Cl or I). In some dimers the Pt(III) atoms are not equivalent with the following chromophores: PtN_4Pt□ + PtN_2S_2Pt□; PtN_4Pt□ + $PtCl_3O_2Pt$; PtN_4OPt□ + PtO_2N_2ClPt□; PtN_4ClPt□+ PtO_2N_2ClPt□; PtN_4OPt□+ PtN_3O_2Pt□; PtN_3ClSPt□+ PtS_3O_2Pt□; PtN_2S_2ClPt□+ PtS_3N_2Pt□; and PtN_2S_2BrPt + PtS_3N_2Pt□ Usually the unidentate X and P atom occupied axial positions. The Pt-Pt bond distance range from 2.390(1) Å to 2.766(1) Å (ave. 2.585 Å).

3.6. Pt(IV) Dimers

There are six colored Pt(IV) dimers. In three of them a pair of a single atoms serve as a double bridge between equivalent moieties, between $Pt(OH)_2(EtNH_2)_2$ are two μ-OH groups; between $Pt(NH_3)_4$ are two μ-NH_2 groups, and between $Pt(NO)Cl_3$ are two μ-Cl atoms. The Pt(IV)-Pt(IV) separations and the mean Pt-X-Pt bridge angles are: 3.090(1) Å and 98.2(3)° (X=OH), 3.85 Å and 100° (NH_2); 3.815(1) Å and 96.5(3)° (X=Cl). In orange yellow dimer two $PtBr_3$ units are triple bridged by three bromine atoms. The Pt-Pt separation is 3.230(2) Å and the mean Pt-Br-Pt bridge angle 79.6°.

In another Pt(IV) dimer between two PtH(η^2-$Ph_2PCH_2PPh_2$) moieties a pair of SiHPh ligands serve as double bridge via Si atoms with the mean Pt-Si-Pt bridge angle of 105.1(2)° and Pt-Pt separation of 3.787 Å. In addition the disilene (PhHSiSiHPh) bridges the Pt(IV) centres along the Pt-Pt vector with a Si-Si single bond of 2.385(6) Å.

In orange yellow two equivalent Pt(η^3-tac)Cl (tac=1,4,7-triazacyclononame ([9]ane N_3)) moieties are double bridged by two peroxo groups in the manner Pt-$(O\text{-}O)_2$-Pt. Each Pt(IV) atom in these Pt(IV) dimers are hexa-coordinated with the inner coordination spheres: PtO_4N_2, PtN_6, $PtCl_5N$, $PtBr_6$, $PtSi_3P_2H$ and PtN_3O_2Cl. The mean Pt-L bond distance elongated in the order: 1.997 Å (μ-η^2-OO) < 2.05 Å (μ-NH_2) < 2.054 Å (ter-NL) < 2.06 Å (NO) < 2.20 Å (NH_3) < 2.338 Å (bi-PL) < 2.386 Å (μ-SiL) < 2.420 Å (Br) < 2.433 Å (μ-η^2-Si_2L) < 2.523 Å (μ-Br) < 2.55 Å (μ-Cl).

3.7. Mixed-Valence Pt Dimers

Two colored derivatives, orange and yellow, contain a Pt(0) and a Pt(II) atom. In the former examples, the two metal atoms are bridged by one diethylphosphide group and by one arm of the $N(CH_2CH_2PPh_2)_3$ ligand, using two P atoms and N atom to bind the Pt(II) and the third P atom binds to Pt(0). Around the Pt(II) centre an N atom is positioned approximately trans to the phosphide group with N-Pt-P angle of 161.8(4)°. Around the Pt(0) centre the PPh_3 is trans to the Pt(II) atom with Pt(II)-Pt(0)-P angle of 157.5(1)°. The Pt(0)-Pt(II) bond distance is 2.824(1) Å and Pt(0)-P-Pt(II) bridge angle is 78.0(2)°. The inner coordination spheres are $Pt(0)P_3Pt(II)$ and $Pt(II)P_3NPt(0)$. A yellow Pt(0)/Pt(II) dimer has three $F_2PN(Me)PF_2$ ligands as triple bridges $Pt(PNP)_3Pt$ with a Pt-Pt bond distance of 2.8911(4) Å. The Pt(0)-P bond distance of 2.187(2) Å is about 0.06 Å shorter than the Pt(II)-P distance (2.247(2) Å). The Pt(0) is four- PtP_3Pt and Pt(II) is five-PtP_4Pt coordinated, because of the PPh_3 ligand which bind to the Pt(II). The structure of Pt(II)-Pt(III) dimer consists of a complex cation $[Pt_2(dtf)_4]^+$ (dtf=N,N′-di-p-tolylformamidinate) in which four bidentate dtf ligands quadruple bridge two non-equivalent Pt atoms in the manner Pt(II)-$(NCN)_4$-Pt(III) with a Pt-Pt bond (2.5304(6) Å).

There are five dimers which contain Pt(II) and Pt(IV) metal centres. In yellow [ClPt(II)(μ-mp)$_2$Pt(IV)Cl_3] (mp=3-mercaptomethyl-1-methyl piperidinate), two Pt atoms are linked by two bridging S atoms and form a central $Pt(S)_2Pt$ core with the Pt(II)-Pt(IV) distance of 3.331(1) Å and the mean Pt-S-Pt bridge angle of 92.6(1)°. In addition each mp ligand is coordinated via N atom to each Pt centre. Metric parameters around planar $Pt(II)S_2NCl$ are significantly shorter than those corresponding to the equatorial plane of octahedral $Pt(IV)Cl_3S_2N$, with the values Pt(II)-μS and Pt(II)-N vs Pt (IV)-μS and Pt(IV)-N of: 2.269 and 2.109 Å vs 2.343 and 2.230 Å. However, the N-Pt-S bite angle of 94.3(2)° of Pt(II) is about 1.6° wider than that of Pt(IV) at 92.7(2)°.

Another three bronze Pt(II)/Pt(IV) dimers contain $[I_2Pt(II)(\mu\text{-}I)_2Pt(IV)I_4]^{-2}$ anions, which are isostructural. Two iodine atoms serve as double bridge between respective units with the mean Pt-I-Pt bridge angles in the central Pt_2I_2 ring of 95° and Pt-Pt separation of 3.866(4) Å. The parameters around planar $(Pt(II)I_4)$ centre are significantly shorter than those corresponding to the equatorial plane of the octahedral $Pt(IV)I_6$ centre (Pt(II)-μI/I vs Pt(IV)-μI/I, (mean values), 2.608/2.590 vs 2.644/2.648 Å).

In another dimer, two $Pt(II)(PEt_3)_2$ and $Pt(IV)(PEt_3)_2$ units are bridged by two bidentate 1,2-disilylbenzene ligands via Si donor atoms. One of the Si atoms of each ligand binds both Pt atoms creating $PtSi_2Pt$ ring system with the mean Pt(II)-Si-Pt(IV) bridge angle of 111.2(4)°. Another Si atom binds only to the Pt(IV) and forms two five-membered metallocycles with mean Si-Pt(IV)-Si bite angle of 70.0(4)°. The Pt(II) adopts a distorted square planar (PtP_2Si_2) and the Pt(IV) a pseudooctahedral $(PtSi_4P_2)$ arrangement.

4. Trimeric Pt Coordination Complexes

4.1. Pt(0) Trimers

There are ten colored, mostly orange and yellow with one green and brown, Pt(0) trimers. Their structures have an approximate isosceles triangle of Pt(0) atoms with Pt(0)-Pt(0) bond distances range from 2.618(1) to 2.887(1) Å. The maximum deviation of Pt-Pt-Pt triangle angle from 60° is 2.8° (range 0.1 – 2.8°, ave. 0.95°). In five trimers each edge of the triangle is bridged by μ-SO_2 groups, in another three trimers by μ-CNxyl group and remaining two by μ-PL ligands. In the last trimer, two edges are bridged by μ-SO_2 groups and remaining one by μ-Br atom. The mean Pt(0)-X-Pt(0) bridge angle opens in the order of X: 69.3° (Br) < 71.0° (P) < 75.1° (S) < 79.8° (C) which corresponds with decreasing covalent radii of X atom. Each Pt(0) atom is also terminal coordinated by PL (PPh_3 or Pcy_3) or isocyanide ligands. The Pt(0) atoms in some trimers have equivalent inner coordination spheres: PtC_3Pt_2, PtS_2PPt_2, P tC_2PPt_2, and in some others non-equivalent: PtS_2PPt_2 (x2) plus $PtS_2P_2Pt_2$; PtC_3P t_2 (x2) plus PtC_2PPt_2; PtC_2PPt_2 (x2) plus PtP_2CPt_2; $PtCSPPt_2$ (x2) plus PtC_2PPt_2; and $PtSPBrPt_2$ (x2) plus PtS_2PPt_2. The mean Pt(0)-L bond distance elongated in the order: 1.92 Å (CNL) < 2.287° (PL) < 2.337 Å (bi-PL) and the mean Pt(0)-L (bridge) bond distance: 2.06 Å (μ-CNL) < 2.260 Å (μ-SO_2) < 2.324 Å (μ-PL) < 2.536 Å (μ-Br).

4.2. Pt(I) Trimers

There are thirteen mostly yellow Pt(I) trimers and in three of them the metal atoms form a triangle with the Pt(I)-Pt(I) bond distances range from 2.565(1) to 2.797(1) Å (ave. 2.66 Å).

The departure of the Pt-Pt-Pt angle from ideal (90°) are 3.7, 2.0 and 2.3°. In the former one edge of the Pt_3 triangle is bridge by μ-H and remaining two by μ-PPh_2 groups. In the remaining two each Pt-Pt edge is linked by one of three μ-P^2-$Ph_2PCH_2PPh_2$ ligands. In another Pt(I) trimer the three Pt(I) atoms are collinear with the two P atoms of two PPh_3 groups forming a five-atom chain with Pt(1)-Pt(2)-Pt(1) angles of 180.0° and Pt(1)-Pt(2)-P of 171.24(8)°. The Pt(2)-Pt(1) distances are equal at 2.6437(4) Å. In the next trimer, the Pt_3 triangle is linked by μ-$Ph_2PCH_2PPh_2$ ligands and a triple bridging S atom. The Pt(1)-Pt(3) distance of 2.597(4) Å is indicative of a single bond, where as the other two Pt-Pt distances (3.574(2) and 3.678(7) Å are clearly non-bonding. An isosceles triangle is found in the other trimer, with two Pt-Pt distances from the triangle apex of 2.683(2) and 2.641(2) Å, but a baseline Pt-Pt separation of 4.01 Å which precludes bonding. Thc Pt-Pt-Pt angle at the apex is 99.73(6)°. The Pt_3 triangle is linked by μ-$Ph_2P(CH_2)_2PPh_2$ ligands.

There are six trimers with an A frame structure. The Pt framework is bridged by two $Ph_2PCH_2PPh_2$ (dppm) ligands. The dppm ligand bridges over the Pt(2) and Pt(2′) atoms with the outer and inner P atoms, and between the Pt(l) and Pt(2) atoms with the other pair of P atoms. The Pt(1)-Pt(2) bond distances range from 2.593(1) to 2.681(1) Å (ave. 2.620 Å) and the interatomic Pt(2)-Pt(2′) distances range from 3.219(2) to 3.304(2) Å (ave. 3.204 Å), implying no metal link in the latter. The Pt(I) trimers crystallized in monoclinic (x6) > triclinic (x5) > orthorhombic (x2) crystal classes. There are two categories of Pt(I)-Pt(I) distances, bond distances which range from 2.565(1) to 2.796(2) Å (ave. 2.642 Å), and non-bond distances range from 3.096 to 4.01 Å (ave. 3.405 Å). The mean Pt-X-Pt bridge angles are: 76.4° (P, 1.06 Å) < 85.2° (S, 1.02 Å) < 100.1° (N, 0.77 Å). The mean Pt(I)-L bond distance elongated in the order: 1.94 Å (CNL) < 2.02 Å (μ-NO) < 2.15 Å (H) < 2.26 Å (bi-PL), 2.26 Å (μ-PPh_2) < 2.285 Å (PL) < 2.32 Å (ter-PL) < 2.34 Å (μ_3-SO_2). The Pt(I) atoms adopts in a triangular array, each Pt(I) atom has a five-coordinate inner sphere including two other Pt atoms (PtX_3Pt_2). In other trimers the Pt(I) atoms are four-(distorted square planar) or with mixed five- and four-coordinated as well.

4.3. Pt(II) Trimers

There are almost forty Pt(II) trimers and their structures are fairly complicated. Three colored trimers are very similar with a cluster core comprised to three Pt(II) atoms in an approximate isosceles triangle. Two equidistant sides at a mean values of 2.54 Å are shorter than the third side (mean 2.60 Å). Each Pt(II) atom adopts a distorted octahedral geometry ($PtO_2N_2Pt_2$). In a yellow derivative the Pt_3 triangle has two equidistant sides at 2.82 Å. Three of the hydride ligands bridge along the sides of the triangle, while the other three (H) are terminal ligands on each P atom. With a PBu^t_3 ligand each the Pt(II) atoms have a distorted octahedral geometry (PtH_3PPt_2). In another two colorless derivatives three Pt(II) atoms are triple bridged by OH or an O atom in propeller arrangement. Each Pt(II) atom is further bridged by consecutively, by two Ph_2PO ligands, forming a five-membered metallocycle with a central O atom. Three PPh_3 in one and $PMePh_2$ in the other one are terminally coordinated to complete the square-planar geometry (PtO_2P_2).

The core structure of $[Pt_3(\mu_3\text{-}X)_2(PMe_2Ph)_6]^{+2}$ (X=O or S) and $[Pt_3(\mu_3\text{-}S)(\eta^2\text{-dppe})_3]^{+2}$ (X=S or Te) consists of three square planar PtX_2P_2 planes sharing two μ_3-X ligands. The central core of Pt_3X_3 is of trigonal bipyramidal geometry with the mean Pt-Pt distances of

2.877 Å (X=O), 3.122 and 3.157 Å (X=S) and 3.456 Å (X=Te). The mean Pt-X bond distance elongated with an increasing covalent radii of X, in the order: 2.083 Å (O, 0.75 Å) < 2.366 Å (S, 1.02 Å) < 2.631 Å (Te, 1.36 Å). A yellow $Pt_3H_3\{\mu\text{-}MeC(CH_2S)_3\}(PPh_3)_3$ has three Pt(II) atom at the vertices of a triangle, whose edges range from 3.485(1) to 3.587(1) Å, held together by sulphur thiolate bridges. Alternating metal and sulphur atoms form a six-membered ring, with a chair conformation. Each Pt(II) atom adopts a distorted square planar environment (PtS_2HP). The structure of an orange Pt(II) trimer comprises a linear Pt_3 chain bridged by two SNCPhNS ligands in which the terminal Pt(1) atoms are related via a crystallographic inversion center at Pt(2). Each Pt(II) atom has planar geometry PtS_2P_2 at each end and PtS_4 at the central metal atom. The Pt(1)-Pt(2) distance is 2.865(1) Å.

The structure of $Pt_3(\mu\text{-}\eta^2\text{-pz})_6$ forms two nine-member ring containing all tree Pt atoms ($-PtN_2-)_3$ with overall symmetry D_{3h}. The Pt(II) atom are each bridged by two pyrazolate ligands with Pt-Pt distances of 3.034(1) and 3.067(1) Å. In another two Pt(II) trimers three bidentate l-methylcytosinate ligands (mec) bridged three Pt(II) atoms, creating a twelve-membered $(-PtNCN-)_3$ ring with Pt-Pt separations of 5.31 Å. Two PMe_3 ligands complete a distorted square planar geometry (PtN_2P_2). In another trimer, three bidentate mec ligands with three bidentate en ligands, which serve as bridges between Pt(II) atoms, form a twenty-one-membered metallocycle with mean Pt-Pt separation of 9.42 Å. In colorless Pt(II) trimer in which three bidentate ethylenediamine and three cyclic oligorolaxane ligands serve as bridge, the mean Pt-Pt separation is 19.48 Å. In three trimers each Pt(II) atom has PtN_4 chromophore.

Two terdentate (3P) 1,3,5-tris(diphenylphosphino)benzene (dppb) ligands serve as bridges between three $PtCl_2$ units, creating two pairs of metallocycles, one eight- and one-ten-membered with Pt-Pt separation of 7.701 and 7.893(1) Å (twice), respectively. In another trimer, three $PtCl_2(PEt_3)$ units are linked by 1,3,5-triazine ligand. In the former each Pt(II) atom has PtN_4 and in the latter $PtCl_2NP$ chromophore.

A yellow derivative contains both trimeric and monomeric units within the same crystal. In the trimer a 9-ethylguanine ligand simultaneously binds all three $Pt(NH_3)_3$ units with different N-donor sites. In the monomer four NH_3 molecules form a square planar geometry (PtN_4), also in the trimer each Pt(II) atom has PtN_4 chromophore. Another trimer contains a central cis- $Pt(NH_3)_2$ units linked by other N-donor sites of terdentate 9-methyladenine (mea) ligands. The central Pt atom is located on a mirror plane, with two adenine ligands arranged in head-to-head fashion (PtN_4). In another trimer two bidentate mea ligands act as bridges between two Pt(9-methylguanine)$(NH_3)_2$ units and a $Pt(MeNH_2)_3$ unit. The two adenine ligands are propeller-twisted by 17.2(4)°. Each Pt(II) atom has PtN_4 chromophore.

In dark orange trimer, three Pt(η^2-dppb) satellite units bonded to a hexadentate cyclotricatechylene via O_6-donor sites, two per Pt atom. The Pt-Pt centres are far apart (9.85 Å) and each adopts PtO_2P_2 chromophore. In another example, two satellite Pt(η^3-trpy) units are bonded to the S-atoms of two 2-aminoethanethiolate ligands, which serves as bridging between the central Pt(II) atom (via S-donor site) and the satellites. In addition the ligand is linked to the central Pt atom also by N-donor sites. The metal atoms have PtN_3S (satellites) and PtN_2S_2 (central) chromophores. The Pt-Pt separations are 3.880 Å (satellite-central) and 4.420 Å (satellite-satellite).

In another example, three $PtCl_2$ units are linked by two terdentate bis(diphenylphosphinomethyl)phenylphosphine ligands via P-donor sites. Each bridge binds to the outer $PtCl_2$ units via end P atoms (P-C-P-C-P) and to the inner unit by the central P atom. The PtP_3C_2 chelate rings display distorted boat conformation with the Pt(II) atoms having $PtCl_2P_2$ chromophore. A yellow derivative has a central hexadentate triquinoline(-3) anion bound via O and N atoms to three $PtCl(PBu_3)$ units. Each Pt(II) has PtONClP chromophore. In another trimer a hexadentate dphpc ligand uses P-donor sites to bind three $PtCl_2$ units in a chelating fashion.

In the series of Pt(II) trimers, the Pt(II) atoms are six-, five- and four- coordinate (most common). The ligands include uni-, bi-, ter-, tetra- and hexadentate types. The mean Pt(II)-L bond distance elongated in the sequence: 2.01 Å (bi-OL) < 2.04 Å (ter-NL) < 2.05 Å (uni-NL) < 2.055 Å (bi-NL) < 2.255 Å (uni-, bi-PL) < 2.26 Å (ter-PL) < 2.315 Å (ter-SL) < 2.46 Å (Br) < 2.58 Å (I). The mean Pt(II)-L (bridge) bond distance elongates in the order: 2.04 Å(OL) < 2.33 Å (Cl) < 2.36 Å (SL) < 2.43 Å (Br) < 2.585 Å (I). The mean Pt(II)-μ_3L bond distance elongates in the order: 2.115 Å (O) < 2.360 Å (S) < 2.625 Å (Te). In general, the bond distance elongates with increasing covalent radius of the coordinating atom. The Pt-Pt distances range from 2.516(1) to 19.476(1) Å. In the trimers which contain a Pt_3 triangle the Pt-Pt bond distances range from 2.516(1) to 2.82(2) Å, with an average value of 2.66 Å. The Pt-L-Pt bridge angles open with decreasing covalent radius of the bridging atom in the order: 85° (Te, 1.36 Å) < 93.5° (Se, 1.16 Å) < 96.0° (Cl, 0.99 Å) < 106.0° (O, 0.75 Å).

4.4. Mixed-Valence Pt Trimers

There are two mixed-valence Pt trimers. Dark brown contains $[Br_4Pt(\mu\text{-}Br)_2Pt(\mu\text{-}Br)_2PtBr_4]^{-2}$ anion in which a central Pt(II) atom is surrounded by four double-bridged Br atoms ($PtBr_4$). Two Pt(IV) atoms are located in satellite positions, bridged to the two Br atoms, with four terminal Br atoms each ($Pt(IV)Br_6$). In black trimer two planar Pt(II)$I_3(NHEt_2)$ units share edges with the central $Pt(IV)I_6$ octahedron. The mean Pt(II)-X (bridge) versus Pt(IV)-X (bridge) distance are: 2.428(4) vs 2.507(4) Å (X=Br); 2.585(4) vs 2.668(4) Å (X=I).

5. Tetrameric Pt Coordination Complexes

5.1. Mixed-Valence and Non-Integral Pt Oxidation States

There are seventeen tetrameric clusters with such oxidation states of Pt. The structure of a yellow cluster consists of a quasi-linear P-Pt_4-P frame with mean P-Pt-P and Pt-Pt-Pt angles of 173.9(2) and 175.3(5)°, respectively. The Pt(0.5)-Pt(0.5) bond distances are 2.655(2) and 2.666(3) Å for the outer and 2.641(2) Å for the inner bond. The four 1,8-diisocyano-p-methane ligands (dmb) bridge the Pt(0.5) atoms forming a catenate. Two of the dmb ligands bridge the first and third Pt atoms and two others bridge the second and fourth. In dark red cluster, $Pt_4H_2(PBu^t_3)_4$, one triangular $\{Pt_3\}$ lies on a crystallographic mirror plane, about which the rest of the molecule is disordered. The Pt(0.5)-Pt(0.5) bond distances range from 2.667(3) to 2.810(3) Å (ave. 2.737 Å).

Structure of dark green cluster consists of well separated $[Pt_4H_2(PBu^t_3)_4]^{+2}$ cation, BF_4^- anions and HBF_4 molecules. Structure of the complex cation is a platened tetrahedron $\{Pt_4\}$

with the PBu^t_3 ligands each arranged to lie approximately trans to one Pt(II)-Pt(I) bond. The Pt_4P_4 core has S_4 symmetry. The mean Pt(I)-Pt(I) bond distance is 2.722 Å and mean Pt-Pt-Pt angle is 61.0°.

There are three bright orange clusters of the composition $[(dmso)_2Cl_2Pt(I)_2(\mu_3\text{-}O)_2Pt(II)_2(dmso)_4]$, with a different solvated molecules and their structures are almost identical. A dimeric $Pt(I)_2Cl_2(dmso)_2$ unit is fused into a dimeric $Pt(II)_2O_2(dmso)_4$ unit by the formation of two μ_3-O bridges. Each μ_3-O bridges one Pt(I) and two Pt(II) atoms, and the Pt_4O_2 moieties forms a distorted cubane-like framework in which two corner O atoms are absent. The O-Pt(I)-Pt(I)-O face deviates significantly from planar with torsion angles of about 28°. The mean Pt(I)-Pt(I) bond distance is 2.584(3) Å. Each Pt atoms four-coordinated: Pt(I)OClSPt and $Pt(II)O_2S_2$.

There are ten colored tetramers contain complex $[(NH_3)_2Pt(\mu\text{-}\eta^2\text{-}L)_2Pt(NH_3)_2]_2^{+n}$ cations (L=2-pyrrolidinate (pyro); (6 examples); 1-methluracilate (meu); 2-pyridonate (pyo); or 3,3-dimethylglutariminate (dmgi); $[NH_3)_2Pt\ (\mu\text{-}\eta^2\text{-}met)_2PtCl_2]_2^{+n}$ (met=1-methylthiminate) and one $[(\eta^2\text{-}en)Pt(\mu\text{-}\eta^2\text{-}pyo)_2Pt(\eta^2\text{-}en)]_2^{+n}$ with a similar structural architecture, consisting of two cis-$Pt(NH_3)_2$ (or $Pt(\eta^2$-en)) unit bridged by two μ-η^2 ligands. Two of them are further linked across a crystallographic inversion centre to form a tetrameric unit. The Pt atoms at the ends of the chain are bonded to two ammine ligands (or ethylenediamine) in a cis- configuration, and to two N atoms of two bridging ligands. The inner Pt atoms have two cis-ammine groups (or ethylenediamine) and two exocyclic (ligand) ring O atoms in their coordination sphere (head-to-head fashion). Apart from the bridging ligands, the two square-planar PtN_4 and PtO_2N_2 moieties are further joined by a Pt-Pt bond. The formal oxidation states of Pt in these tetramers are non-integral with the mean Pt-Pt bond distance elongated in the order: 2.699 Å (Pt (2.75)-Pt(2.75)) < 2.730 Å (Pt(2.50)-Pt(2.50)) < 2.746 Å (Pt(2.375)-Pt(2.375)) < 2.866 Å (Pt(2.14)-Pt(2.14)) < 2.867 Å (Pt(2.25)-Pt(2.25)). The mean Pt-L bond distance elongated in the sequence: 1.93 Å (bi-CNR) < 2.03 Å (bi-NL) < 2.045 Å (NH_3) < 2.195 Å (SL) < 2.26 Å (PL) < 2.375 Å (Cl).

5.2. Pt(II), Pt(III) and Pt(IV) Tetramers

There are nineteen Pt(II), four Pt(III) and three Pt(IV) tetramers, for which X-ray data are available. Four colored clusters $[Pt(\mu\text{-}\eta^2\text{-}ac)_2]_4$ (tetragonal and monoclinic); $[Pt(\mu\text{-}\eta^2\text{-}ac)(\mu\text{-}\eta^2\text{-}Cl_3ac)]_4$ (tetragonal) and $[Pt(\mu\text{-}\eta^2\text{-}F_3ac)_2]_4$ (triclinic) (ac=acetate) have a structure based on a strongly metal-metal bonded square cluster of four Pt(II) atoms. The Pt-Pt bond lengths range from 2.492(1) to 2.513(1) Å (ave. 2.50 Å). Eight bridging $RCOO^-$ groups are in a syn-syn arrangement around the square such that four groups are in the plane and the other four alternatively above and below the plane. The Pt(II) atoms are thus octahedrally coordinated (PtO_4Pt_2). The eight Pt-O bonds in the cluster plane are rather long with a mean value of 2.15 Å. The Pt-O bonds out of plane have a mean length of 2.01 Å.

An orange cluster contains well separated $[Pt_4(\mu\text{-}\eta^2\text{-}ac)_6py_4]^{+2}$ cation and a PF_6^- anions. Six bridging acetate groups are arranged in a syn-syn arrangement about the four Pt(II) atoms with terminal pyridine ligands. The Pt-Pt bonds in an approximate square array are 2.522(5) and 2.552(3) Å. Each Pt(II) atom has a pseudo-octahedral environment (PtO_3NPt_2). A complex $[Pt_4(\mu\text{-}\eta^2\text{-}ac)_5(\mu\text{-}\eta^2\text{-}N_6L)]^{+3}$ cation in another orange cluster, has an approximate square of Pt(II) atoms with Pt-Pt bonds ranging from 2.5272(7) to 2.5418(7) Å (ave. 2.534

Å). Five bridging acetate groups are arranged around the square, and in addition the hexamine ligand (substituted piperazine), in a bridge-chelate-bridge-chelate-bridge mode and one acetate ligand occupy coordination sites in the plane of the cluster. Each Pt(II) atom adopts a pseudo-octahedral geometry (two PtO_3NPt_2 and two $PtO_2N_2Pt_2$).

In two red clusters a square of Pt_4 atoms are bridged by four acetate groups in a syn-syn arrangement. The mean Pt-Pt bond distance is 2.549 Å. In one cluster a bidentate 2-pyridinecarboxylate, in the other an ethylenediamine, chelate each Pt(II) atom to give pseudo-octahedral geometry, PtO_3NPt_2 and $PtO_2N_2Pt_2$ clusters, respectively.

A complex $[Pt_4H_7(PBu^t_3)_4]^+$ cation in red brown cluster, has a butterfly configuration with a distinct head-to-tail fashion. The Pt-Pt bonds ranges from 2.636(2) to 2.870(2) Å (ave. 2.776 Å). An orange $Pt_4H_8(PPr^i_2Ph)_4$ cluster, reveals a central Pt_4P_4 unit with S_4 symmetry, with two short (2.872 Å) and two long (3.085 Å) Pt-Pt distances. A red violet cluster has a rectangular metal skeleton with long sides (3.235(1) Å) spanned by μ-NO and μ-η^2-NO_2 ligands. The short side (2.889(7) Å) is spanned by pairs of μ-η^2-$OOCCHNMe_2$ ligands. Neighboring molecules, related to a symmetry centre, from a 3.144(1) Å contact with each other without any bridging. A red cluster $[\{(\mu\text{-}\eta^2\text{-ac})_2Pt\}_2(\mu\text{-}\eta^2\text{-ac})(\mu\text{-NO})]_2$ has four Pt(II) atoms in an approximate rectangle of sides 2.944(2) and 3.311(2) Å. Each of the two independent metal atoms is bonded to three O atoms from bridging acetate groups and to one N atom from a bridging nitrosyl group, resulting in square planar coordination (PtO_3N). A similar unabridged neighbor contact (3.330 Å) was found in this case as in the previous one.

In colorless cluster, four $Pt(NH_3)_2$ units are connected by a pair of μ_4-O_2CO groups forming four fused six-membered rings. The Pt-O backbone may be visualized as an eight-membered ring system of approximate S_4 symmetry, with bridging O atoms above and below the Pt_4 plane. Each Pt(II) atom has a square planar geometry (PtO_2N_2). A bright green cluster, has two dimers connected via Pt(II) centres (3.197(3) Å). In each dimer two cis-$Pt(NH_3)_2$ units are bridged by a pair of squarate divalent anions with Pt-Pt distance of 3.162(2) Å. Each Pt(II) atom has square planar geometry (PtO_2N_2), and the complex might be more appropriately considered as a pair of dimeric rather than a tetramer. In pale yellow a $[Pt(\mu\text{-}OH)(\eta^2\text{-en})]_4^+$ cation consisting of an eight-membered ring with approximate S_4 symmetry. Each Pt(II) atom in the hydroxo-bridged tetramer has square planar geometry (PtO_2N_2). The four Pt atoms in the ring are planar with two cross-ring O atoms on one side of the plane and two on the other side. The mean Pt-O-Pt bridge angle is 112.6(9)° and the Pt-Pt separation ranges from 3.321(2) to 3.582(3) Å, which excludes a bond. In another yellow cyclic $[Pt\{\mu\text{-}\eta^3\text{-Ph(O)P CH}_2CH_2P(O)Ph\}]_4$ each Pt(II) atom is chelated by a meso ligand via P atoms and cis-bonded to two phosphoryl O atoms (P=O) originating from separate meso ligands in an approximate square planar geometry (PtO_2P_2). The Pt-Pt separations range from 3.335(ä2) to 5.124(3) Å. In blue complex, a cyclic tetramer is formed by four uracilate monovalent anions and four Pt(II) atoms. Each Pt(II) atom is bonded to N(1) of one ring and N(3) of a second ring, and is also chelated by ethylenediamine to complete an approximate square planar geometry (/PtN_4) with a Pt-Pt separation of 5.86 Å. In another tetramer four Pt(η^2-dppp) (dppp=1,3-bis(diisopropylphospino)propane) units occupy the corners of a macrocyclic square with edge to edge Pt-Pt distance of 11.225(4) Å and diagonal distance of 14.299(2) Å. There are four 4,4′-bipyridyl ligands linking pairs of Pt(II) atoms which adopt a distorted square planar geometry (PtN_2P_2). In $[Pt(\mu\text{-}\eta^2\text{-}C_{18}H_{16}N_2)(\eta^2\text{-en})]_4^{+2}$ ($C_{18}H_{16}N_2$=1,4-

$NC_5H_4CH_2C_6H_4CH_2C_5H_4N$) two boat conformation rings interlock with edge to face aromatic-aromatic contact that probably induces the self-assembly of the catenated structures.

There are four colored clusters which contain the Pt(III) atoms. A yellow tetramer consists of two 2-pyrrolidonate bridged head-to-tail Pt(III) dimers double bridged by OH groups. The 2-pyrrolidinate bridged Pt-Pt distance is 2.5527(7) Å. The inner Pt-Pt distance bridged by OH groups is 3.1578(6) Å. Each Pt(III) atom adopts a pseudo-octahedral structure (PtN_3O_2Pt□and PtO_3N_2Pt□). A dark green complex also have two 2-pyrrolidonate bridged Pt(III) dimers, but this time head-to-head double bridged by NH_2^- ligands. The molecule is centrosymmetric at the mid-point of the central Pt(1)-Pt(1) vector. While the inner Pt-Pt distance bridged by NH_2^- is 3.116(2) Å, the Pt-Pt distance bridged by 2-pyrrolidonate is 2.608(1) Å. Each Pt(III) atom has a pseudo-octahedral geometry (PtN_4OPt and PtN_3O_2Pt□).

Two red Pt(III) tetramers are isomorphous. Two [$Pt_2(\mu$-η^2-mpy$)_4X$] (mpy=2-mercatopyridine; X=Cl or Br) units are linked by an end-to-end S_4^{-2} group. Four heterobidentate mpy ligands via N and S donor sites, bring two Pt(III) atoms within 2.557(2) Å (X=Cl) and 2.556(2) Å (X=Br). Each Pt(III) atom adopts a pseudo-octahedral geometry (PtN_2S_2XPt□and PtS_3N_2Pt□). All Pt(III) tetramers are built up by a pair of dimers. In the dimers, the Pt-Pt bonds range from 2.5527(7) to 2.608(1) Å (ave. 2.568 Å). The separation between the dimer units are over 3.00 Å.

Three black tetramers contain Pt(IV) atoms and are isostructural, $[Pt_4(\mu_3\text{-}S)_4.(\mu\text{-}\eta^2\text{-}S_3)_6]^{-4}$ (cubic), $Pt_4(\mu_3\text{-}Te)_4(\mu\text{-}\eta^2\text{-}Te_3)_6]^{-4}$ (cubic and tetragonal). Each of the anion possesses a Pt_4S_4 or Pt_4Te_4 cubane core and six S_3^{-2} or Te_3^{-2} chains bridging Pt(IV) atoms on the cubane faces. Each Pt(IV) atom is in a distorted octahedral environment, being coordinated by three S or Te atoms in the cubane framework, and by three S or Te atoms in chains that bridge to the other three platinum atoms. The mean Pt-Pt distance is 3.478(4) Å in the Pt_4S_4 core and 3.910(2) Å in the Pt_4Te_4 core. The mean Pt-X-Pt angles in the cube are 95.3(6)° (X=S) and 95.68(6)° (X=Te). The pseudo Pt_2X_3 five-membered rings have an envelope conformation. While the mean Pt-μ_3S bond distance of 2.40(2) Å is about 0.12 Å longer than that of a Pt-S bond (2.28(2) Å), the mean Pt-μ_3Te bond distance of 2.637(2) Å is about 0.041 Å shorter than that of a Pt-Te bond (2.678(2) Å).

6. Penta-to Octameric Pt Coordination Complexes

Two orange examples contain five Pt(0) atoms, and both are based on edge- bridged tetrahedral cores, with three bridging and seven terminal ligands. In one cluster the three S-bridging ligands are SO_2 molecules and the seven terminal ligands are 2,6-dimethylpehynlisocyanide (CNxyl) via C-atom. The Pt-Pt distances range from 2.6956(9) to 3.0089(9) Å (ave. 2.804 Å). In the other cluster, the bridging ligands are two SO_2 molecules and one CNxyl, with seven of the latter as terminal ligands. Here the Pt-Pt distances range from 2.689(1) to 2.843(1) Å (ave. 2.780 Å).

Two hexameric derivatives contain Pt(II) atoms. In a yellow example, the six Pt(II) atoms are linked by a double-bridging thiolate S atom into three fused rings, a central eight-membered Pt_4S_4 ring with a chair-like conformation, and two outer six-membered Pt_3S_3 rings, each sharing Pt_2S edge. All eight 2-aminoethanethiolate ligands (aet) are both chelating and bridging. Each Pt(II) atom adopts a distorted square planar geometry (PtS_3N and PtS_2N_2). The

Pt-Pt distances are 3.672(1), 3.751(1) and 3.624(1) Å within the six-membered ring, with the shortest value of 3.507(1) Å being across the inversion centre of the eight-membered ring. A colorless derivative contains a cyclic cation in which six cis-$Pt(PMe_3)_2$ units are symmetrically bridged by diprotonated 4-methylguanine ligands via their N(1) and N(7) atoms. The hexamer, with S_3 symmetry, has the purine rings located alternatively above and below the mean plane of the metal atoms at an average angle of 50°. Each Pt(II) atom has a distorted square planar geometry (PtN_2P_2).

A dark brown $Pt_7(CNxyl)_{12}$ is the only example of a heptamer. The molecule contains a Pt(0) in a distorted trigonal bipyramidal array, with two extra Pt atoms added in bridging positions between one basal Pt and each apical Pt atom. The two Pt-Pt bonds on one edge with a bridging Pt atom have distance 2.733(5) and 2.683(4) Å. Thc remaining Pt-Pt distances can be divided into the two groups with the mean values of 2.682 and 3.004 Å (each of 3 values).

There are six octameric derivatives. In dark brown derivative, two $\{Pt_4(ac)_4(dmgH)(dmgH_2)\}_2$ tetramers are bridged by the η^2-$dmgH_2$ (dmg= dimethylglyoximate) ligand which has a equal C_2 axis at the centre of the tetramers. The tetrameric Pt(II) core consists of two units, a triangular Pt_3 cluster and a square planar Pt monomer, with not Pt-Pt bond between the units. The triangular core is an isosceles triangle with Pt-Pt bond ranging from 2.529(1) to 2.678(1) Å (ave. 2.582 Å). The basic feature of the $[\{NH_3)_2Pt\}_4(\mu$-η^2-uracilato$)_4\{Pt(\eta^2$-en$)_4\}_4]^{+8}$ and $[\{(\eta^2$-en$)Pt\}_8(\mu$-η^2-uracilato$)_4]^{+8}$ cations is the 1,3-alternate arrangement of the four uracil nucleobases. The Pt(en) units linking two N sites display normal geometry, as do the cis-$Pt(NH_3)_2$ or Pt(en) moieties . The Pt-Pt distances range from 2.998 to 8.31 Å. Each Pt(II) atoms adopt a PtN_4 chromophore.

In remaining three octamers the Pt atoms have non-integral oxidation states. In two of these each $[(NH_3)_2Pt(\mu$-η^2-faca$)_2Pt(NH_3)_2]_4{}^{+n}$ (n=2.165 or 2.5) cation is made up of a zig-zag chain of Pt atoms, formed by the tetramerization of four 2-fluoroacetamide bridged, head-to-head, dimers. The Pt-Pt distances in the first cation, where the Pt oxidation state is +2.8, range from 2.835(2) to 2.979(2) Å (ave. 2.928 Å). In the second cation with Pt(2.25), the values range from 2.778(1) to 2.934(1) Å (ave. 2.873 Å). A green octamer is made up of two tetramers $[Pt(2.25)_2(NH_3)_4(\mu$-η^2-pyro$)_2]_2$ and $[Pt(2.5)_2(NH_3)_4(\mu$-η^2-pyro$)(NO_3)]_2$ cations. In each tetramer the 2-pyrrolidinate ligands are bridged in head-to-head fashion. The mean Pt-Pt□bond distances are 2.863(1) Å for Pt(2.25) and 2.769(1) Å for Pt(2.5).

CONCLUSION

In this Chapter is briefly outlined and analyzed crystallographic and structural parameters of platinum coordination complexes. There are almost two thousand and five hundred such complexes from uni- to octameric, testifying to the rich diversity of the platinum chemistry. In this platinum is found in, the oxidation state of zero, +1, +2, +3 and +4, and also non-integral states of +0.5, +2.08, +2.14, +2.25, +2.375, +2.5 and +2.75. The most predominant state is Pt(II) with a square planar geometry and differing degrees of distortion. In lower oxidation states (below +2) the Pt atom is mostly four coordinated (usually square planar), with some three coordinated (mostly of Y shape) and two coordinated (mostly linear). In the higher oxidation state (above +2) the Pt atom is mostly six coordinated (tetragonal bipyramidal most common), with some five coordinated (mostly trigonal bipyramidal).

Several relationships are found between the Pt-Pt distance, Pt-L-Pt bridge angle and type of bridging and also between the Pt-L, L-Pt-L and covalent radius of the donor atom. These have been discussed where relevant in each section. There is a wide variability of ligand types from uni- to hexadentate. The Pt(II)-L bond distances tend to be shorter than those of Pt(III). This may be due to the differences in coordination number mentioned above, six for the latter and four for the former. In general the Pt-L bond distance elongates with increasing covalent radius of the donor atoms. In general, the bond distances trans to P are longer than their mean values indicating a greater trans influence of P over O, N, Cl or S atoms. However, the Pt-P bond distance shows elongate when the P atom is trans to the hydride (H) ligand, indicating the greater trans influence of the hydride. The Pt-L (terminal) bond distances are somewhat shorter than the corresponding Pt-L (bridge) bonds. The multidentate ligands form various metallocycles. The effects of both electronic and steric factors are operative and influence on the L-Pt-L bite angles.

The chemistry of platinum coordination complexes is rich not only of number for which X-ray data are known, but there is also wide variety of Pt oxidation states as was already mentioned. In many n-mer complexes Pt-Pt distances are below 3.0 Å which can be classifying as bond. The mean Pt-Pt bond distance elongates in the sequence: 2.472 Å (Pt(III)-Pt(III)) < 2.530 Å (Pt(II)-Pt(III)) < 2.546 Å (Pt(II)-Pt(II)) < 2.574 Å (Pt(I)-Pt(I)) < 2.640 Å (Pt(0.5)-Pt(0.5)) < 2.668 Å (Pt(2.5)-Pt(2.5)) < 2.672 Å (Pt(0)-Pt(0)) < 2.700 Å (Pt(2.75)-Pt(2.75)) < 2.724 Å (Pt(2.375)-Pt(2.375)) < 2.775 Å (Pt(2.25)-Pt(2.25)) < 2.825 Å (Pt(0)-Pt(II)) < 2.835 Å (Pt(2.08)-Pt(2.08)) < 2.840 Å (Pt(2.14)-Pt(2.14)). In general Pt(IV) complexes do not involve metal-metal bonding even mixed with Pt atoms in lower oxidation state.

This chapter represents an overview of almost two thousand and five hundred platinum coordination compounds. Correlations are pointed out that may assist in prediction of the properties and behavior of platinum complexes in a variety of industrial, biological and environmental situations.

In addition, it must be noted, that about 10% of platinum, coordination complexes exist as isomers. Detailed analyzed of their structural parameters has been outline by us [8]. In [8] has focused on stereoisomers of platinum coordination complexes. There are five types of isomerism, distortion (65%), cis-trans (30%), fac-mer, mixed isomers and ligand isomers. Despite the importance of cis-trans geometry in the chemistry of Pt(II) complexes compared to other transition metal systems, within platinum chemistry distortion isomerization is far more common. Such a dominance of this form of isomerism is also seen in the chemistry of copper [9].

REFERENCES

[1] Rosenberger, B., van Camp, L., Transco, Y.E. and Mansoour, V.H., *Nature,* 222(1969) 385.

[2] Holloway, C.E. and Melník, M., *Rev. Inorg. Chem.,* 22 (2002) 163; and ref's cited therein.

[3] Holloway, C.E. and Melník, M., *Rev. Inorg. Chem.,* 23 (2003) 125; and ref□s cited therein.

[4] Holloway, C.E. and Melník, M., *Rev. Inorg. Chem.,* 24 (2004) 135; and ref□s cited therein.

[5] Holloway, C.E. and Melník, M., *Rev. Inorg. Chem.,* 24 (2004) 301; and ref□s cited therein.

[6] Holloway, C.E. and Melník, M., *Rev. Inorg. Chem.,* 25 (2005) 93; and ref□s cited. therein.

[7] Holloway, C.E. and Melník, M., *Main Group Met. Chem.,* 29 (2006) 73; and ref□s cited therein.

[8] Melník, M. and Holloway, C.E., *Coord. Chem. Rev.,* 250 (2006) 2261; and ref□s cited therein.

[9] Melník, M., *Coord. Chem. Rev.,* 47 (1982) 239; and ref□s cited therein.

In: Platinum: Compounds, Production and Applications
Editors: L.Varennikov and E. Yedemsky
ISBN: 978-1-62257-939-6

Chapter 2

THE PROBLEMS OF GEOCHEMICAL "ANTAGONISM" BETWEEN REE AND PGE IN MAFIC AND ULTRAMAFIC ROCKS AND THEIR PETROGENETIC SIGNIFICANCE

Lesnov Felix Petrovich
Russian Academy of Sciences, Siberian Branch,
Sobolev Institute of Geology and Mineralogy,
Department of Geodinamic and Magmatism, IGM SB RAS, Russia

It is known that REE began to be used as geochemical indicators for the diagnosis and classification of magmatic, metamorphic and metasomatic rocks and their minerals, as well as for reconstructions of the processes of their formation as developed the analytical methods for determining its quite low concentrations in the analyzed samples. At present, the analysis of REE in geological objects became a rather routine procedure even while studying the ultramafic and mafic rocks and minerals depleted by them. Due to this reason, the rare contemporary publications on the problems of Geochemistry, Mineralogy, Petrology of magmatic rocks and geodynamics do not use the results of REE analysis by one of the existing analytical methods. Approximately the same trend is observed with respect to platinum group elements (PGE), the determination of relatively low concentrations of which became possible with the introduction of several new analytical techniques including ICP-MS method.

In recent years, some studies were executed aiming to identify relationships between REE and PGE in rocks of ultramafic and mafic composition [Lazarenkov et al., 1989; Naldrett et al., 1994; Roman'ko et al., 1994; Shengrong et al., 1994; Lesnov et al., 2000, 2001, 2004a, 2005a,b; Pearson, Woodland, 2000; Lesnov & Oydup, 2002; Lesnov, 2003a, b, c; Schmidt et al., 2003; Lesnov & Balykin, 2005; Zhou et al., 2005; Karipi et al., 2006; Day et al., 2008; Lesnov, 2006, 2009; and other]. Despite the limited number of analyses performed, the results of these studies have shown that comprehensive data on the contents of REE and PGE provide an opportunity to get some additional information that can be used in carrying out geochemical discrimination and systematization of ultramafic and mafic rocks, as well as massifs and complexes composed of them.

In this paper, the general aspects of relations between REE and PGE are discussed on the basis of the geological materials and more then 200 analyses for REE and PGE that were obtained on the 22 mafic-ultramafic massifs and complexes of various geneses that are located in regions of Russia, Mongolia, Kazakhstan, China, Germany, Greece, Vietnam, South Africa, Canada, Australia, Colombia, and others.

MASSIFS OF TUVA, RUSSIA

For a long time we have been studying the problems of petrology of mafic-ultramafic massifs located in the territory of Tuva, including those associated with distribution of REE and PGE impurities in their rocks. At the initial stage the determination of REE was performed by using INAA method and the determination of PGE – by atomic absorption method [Lesnov & Tsimbalist, 1983; Lesnov et al., 2000, 2001; Lesnov & Oydup, 2002; Lesnov, 2003a, b; Lesnov et al., 2004; Lesnov, 2009]. As a result, in the studied rocks we identified a trend of inverse relation between the contents of REE and PGE. Rock samples for studies were selected from the massifs located in the central (Mazhalyksky, Kalbagdagsky, Karashatsky, Maysky), eastern (Bulkinsky) and western (Khobseksky, Birdagsky, Khayalygsky) areas of the territory. These massifs are characterized by structural-tectonic position, size, internal structure, petrographic composition and quantitative correlations between ultramafic and mafic rocks composing them. In these massifs there were more or less clearly marked evidences of an earlier formation of ultramafic rocks in relation to gabbros, which allowed to considering these massifs as polychronic and polygenic plutonic associations [Lesnov, 1986]. The model of the formation of polygenic mafic-ultramafic massifs considered in the work stated is based on numerous field and petrographic observations made on such objects. According to this model, the tectonic intrusion of ultramafic restite protrusions of various sizes, morphologies, and structural positions preceded the intrusion of gabbro intrusions that are spatially brought together with them, which occurred through the same faults. The ultramafic restites that compose these massifs are usually represented by harzburgites, lherzolites and dunites, as well as serpentinized derivatives thereof. Gabbro intrusions that are spatially brought together with the intrusions, are predominantly composed of gabbro and gabbronorites, rarely containing olivine. Along the contacts of gabbro intrusions with protrusions of ultramafites intruded by them there are often localized contact-reaction zones of different thickness, structures and petrographic composition. The zones are composed of a various hybrid ultramafic and gabbro rocks, including plagioclase-bearing harzburgites, plagioclase-bearing lherzolites, plagioclase-bearing dunites, wehrlites, plagioclase-bearing wehrlites, clinopyroxenites and websterites, their plagioclase-bearing varieties, as well as widely ranging in quantitative-mineral composition olivine gabbro and gabbronorites, which in many cases have a banded texture. Let us concentrate upon a brief review of geological and petrographic features of these massifs, as well as on available data on the distribution of REE and PGE in the rocks composing them, which is based not only on my own observations but involving some of the data from previously published works [Volokhov et al., 1972; Poliakov & Bognibov, 1979; Poliakov et al., 1984; Mongush et al., 1999; and others].

Kalbagdagsky massif with dimensions of 1.5 x 6 km. is exposed over an area of 9 sq. km. Its marginal zones are almost everywhere concealed beneath unconsolidated deposits, with the exception of the south-eastern flank, where the intrusive contact of the massif with the Lower Cambrian volcanic strata is uncovered. The massif is predominantly composed of olivine gabbro, to a lesser extent - of gabbronorites and gabbro, but all of these rocks have good preservation properties. Among the gabbro there are occurring stripe-like bodies of different length, which are composed of plagioclase-bearing wehrlites, plagioclase-bearing clinopyroxenites and olivine clinopyroxenites sometimes containing orthopyroxene impurity. The massif is injected with anorthosite veins containing an impurity of clinopyroxene and olivine.

Mazhalyksky massif is exposed over an area of 1 sq. km. and the dominant part is covered by unconsolidated deposits. It occurs among the Lower Cambrian terrigenous-volcanic formations, together with which it is injected with granite dikes. Uncovered part of the massif is composed of serpentinized harzburgites, dunites, and plagioclase-bearing wehrlites with an impurity of orthopyroxene, olivine clinopyroxenites, clinopyroxenites, troctolites, subordinate amounts of olivine gabbro, gabbro, gabbronorites, among which there are gabbro-anorthosites veins. Plagioclase-bearing wehrlites and pyroxenites contain unevenly distributed and ranging in size and shape porphyroblastic- and pocket-isolations of plagioclase. According to Borodina et al. [2004], the type of the massif is stratiform intrusion, which is formed as a result of intrachamber differentiation of mafic melts. However, according to unconfirmed reports, among gabbros there are occurring certain xenoliths of ultramafites, which suggests their earlier formation in relation to gabbro intrusion. Therefore, we believe that this massif is a polygenic body consisting of an ancient protrusive block of ultramafic restites, a later intrusion of gabbros and various transformation products of ultramafic restites under the influence of mafic melts. According to the analysis of amphiboles from gabbro-norites using Ar-Ar isotopic method the age of Mazhalyksky massif amounted to 484.2 ± 2.3 Ma [Borodina et al., 2004]. According to the study of zircon from the gabbro-anorthosite dikes cross-cutting rocks of the massif using U-Pb isotopic method, the age of gabbro-anorthosites amounted to 478 ± 1.4 Ma [Sal'nikova et al., 2004].

Karaossky massif, which is a part of Kaa-Khem ophiolite belt, is a small xenogeny block that lies among the rocks of granitoid pluton that is intruding metamorphic formations of the age ranging from the Late Proterozoic to Early Cambrian [Rudnev et al., 2006]. This massif is composed of gabbronorites and subordinate olivine gabbronorites, among which there are occurring isolations of websterites and wehrlites.

Maysky massif located on the right bank of Kaa-Khem river is exposing an area of about 40 sq. km. and is divided by faults into several blocks that occur among metaporphyrites and carbonate deposits. It is composed of leucocratic gabbronorites, amphibole gabbros, gabbronorites, olivine gabbros, websterites and anorthosites. Influenced by the close located Early Middle Paleozoic granitoid pluton, the rocks of the massif underwent an intensive amphibolization, potash-feldspatization, silicification, epidotization and chloritization.

Rocks from the other studied massifs located in the territory of Tuva (Bulkinsky, Irbiteysky, Karashatsky, Khobseksky, Birdagsky and Khayalygsky) are presented by plagioclase-bearing lherzolites, wehrlites, plagioclase-bearing wehrlites, plagioclase-bearing websterites, olivine clinopyroxenites, leucocratic troctolites, olivine gabbros, gabbros, gabbronorites, gabbro anorthosites, hornblende-bearing gabbros and hornblendites. REE content determination in them was carried out by using tool neutron activation method and ICP MS method. Almost all the rocks of the studied massifs are significantly depleted by REE, except for some samples that are anomalously enriched with LREE. By the general level of REE accumulation the entire collection of analyzed rock samples can be divided into two groups: 1) depleted, and 2) moderately depleted. The first group consists of rocks from Mazhalyksky, Kalbagdagsky, Karashatsky, Khobseksky and Irbiteysky massifs, the total REE contents of which range from 1.4 to 14 ppm (Table 1). The second group includes rocks that compose Karaossky and Maysky massifs, their total REE contents are slightly higher - 16 to 28 ppm. For most samples of both groups the chondrite-normalized LREE contents are slightly higher than HREE contents and, respectively, their REE patterns have a flat negative slope. Among the rocks of the first group the minimum REE contents are defined in the rocks of Kalbagdagsky massif. It is assumed that those melts, from which the gabbros of Kalbagdagsky and Mazhalyksky massifs crystallized, were more depleted with respect to REE compared to the melts from which the rocks of Karaossky and Maysky massifs crystallized.

Estimations of Pd, Pt and Rh in the rocks of these massifs were originally performed using atomic absorption method (Table 2). For the entire studied collection the total PGE contents range from 29 to 112 ppb. Moreover, the contents of Rh are significantly lower than the contents of Pd and Pt and, in some cases, the content of this element was below the limit of detection by atomic absorption method. In the rocks of Kalbagdagsky, Karashatsky and Khobseksky massifs the contents of Pd, Pt and Rh are generally slightly higher than in rocks from the rest of the massifs studied. In the rocks of Mazhalyksky, Kalbagdagsky and Karashatsky massifs Pd/Pt parameter varies in the range from 1 to 5.6, and in the rocks of Maysky massif this value does not exceed 1.

The observed limited data on the contents of REE and PGE in the rocks of mafic-ultramafic massifs of Tuva allow us to consider as a rough approximation the question of relations between elements of these groups. As can be seen on the presented graphs, the contents of Pd, Pt and Rh in these rocks are generally increasing as the content of La and Ce is diminishing. This fact gives ground to suggest that in these rocks there is an inverse relation between the contents of REE and PGE (Figure 1).

To further examine the relationship between REE and PGE in rocks of Tuva massifs, a part of the samples was studied by ICP-MS method (Figure 2). Development of analytical geochemistry of PGE long time hampered by a number of objective reasons including: 1) lowering the detection limits of these elements that can be achieved using the available methodology, and 2) increasing the requirements on the level of contamination associated with the use of insufficiently pure reagents, water and glassware, and 3) incomplete decomposition of microparticles of platinum minerals in sample preparation, and 4) loss of Os at the volatile phase in sample preparation, and some other reasons.

In addition to the rocks of the massifs described above, determination of REE and PGE by ICP-MS method was performed on samples from *Bulkinsky massif,* which is located in the extreme north-eastern part of Tuva (Ergak-Targak-Tayga ridge, Western Sayan) and is structurally confined to junction of Kurtushibinsky and North-Sayan ophiolite belts [Mongush et al., 2005]. Its elliptical body is exposing an area of about 100 sq. km. with the

Table 1. Rare earth element composition of rocks from some massifs in Central Tuva (Russia) (ppm)

Sample's number	Rocks	La	Ce	Nd	Sm	Eu	Gd	Tb	Tm	Yb	Lu	Сумма РЗЭ	(La/Yb)*n*
Mazhaliksky massifs													
L-106	Wehrlite	0,12	0,40	0,40	0,10	0,04	0,15	0,02	0,02	0,10	0,02	1,37	0,81
L -101a	-"-	0,45	1,20	1,00	0,29	0,14	0,44	0,06	0,04	0,20	0,04	3,86	1,52
Mazh-35-5	-"-	1,00	3,30	3,10	1,10	0,32	0,80	0,16	0,08	0,51	0,07	10,44	1,32
L -104A	Wehrlite Pl-bearing	0,89	3,60	3,00	1,27	0,44	1,39	0,21	0,12	0,64	0,10	11,66	0,94
Mazh -30-4	-"-	1,30	4,20	4,10	1,31	0,48	1,60	0,20	0,12	0,68	0,10	14,09	1,29
Karaossky massifs													
M-35-95	Clinopyroxenite	2,30	7,30	7,70	2,97	0,88	3,00	0,49	0,21	1,37	0,21	26,43	1,13
M-26-98	Websterite Pl-bearing	2,90	7,94	7,40	2,60	0,62	2,60	0,37	0,20	1,16	0,18	25,97	1,69
M -36-98	Gabbro-norite	1,60	5,00	4,50	1,34	0,47	1,60	0,21	0,12	0,78	0,12	15,74	1,38
M -25-97	Amphibole gabbro	3,40	7,30	6,00	2,11	1,02	2,10	0,33	0,18	1,10	0,18	23,72	2,09
Kalbagdagsky massif													
Kal-3-3	Wehrlite	0,77	2,00	1,20	0,35	0,16	1,29	0,11	0,06	0,36	0,06	6,36	1,44
Kal-6-3	-"-	0,64	1,30	0,80	0,32	0,15	0,50	0,06	0,04	0,29	0,05	4,15	1,49
Kal-8-1	-"-	0,97	2,20	1,90	0,49	0,22	0,61	0,13	0,08	0,56	0,09	7,25	1,17
Kal-9-2	-"-	1,54	4,10	2,80	1,00	0,39	1,66	0,24	0,16	1,00	0,16	13,05	1,04
Kal-5-2	Wehrlite Pl-bearing	0,90	2,50	2,00	0,68	0,25	0,80	0,14	0,10	0,69	0,11	8,7	0,88
Kal-3-2	Olivine clinopyroxenite	0,70	1,90	2,10	0,82	0,31	1,10	0,17	0,11	0,70	0,10	8,01	0,67
Kal -10-3	Olivine gabbro	0,50	1,10	0,70	0,38	0,18	0,50	0,08	0,06	0,40	0,07	3,97	0,84
Kal -12-4	Gabbro-norite	0,70	1,20	0,90	0,34	0,19	0,40	0,06	0,04	0,32	0,06	4,21	1,48
Karashatsky massif													
L-50	Websterite Amph-bearing	0,46	1,40	1,10	0,47	0,22	0,89	0,15	0,10	0,72	0,12	5,63	0,43
L -37	Websterite Pl-bearing	0,30	0,70	0,40	0,10	0,06	0,15	0,03	0,02	0,09	0,02	1,87	2,25
L -48	Clinopyroxenite Amph-bearing	0,56	1,40	1,00	0,30	0,10	0,44	0,08	0,05	0,35	0,06	4,34	1,08
L -36	Olivine gabbro	0,15	0,50	0,40	0,13	0,08	0,22	0,04	0,03	0,20	0,03	1,79	0,50
L -44	Gabbro-norite	0,30	0,90	0,60	0,25	0,24	0,43	0,08	0,08	0,51	0,10	3,50	0,39
L -45	-"-	0,20	0,60	0,70	0,30	0,14	0,67	0,10	0,09	0,61	0,10	3,50	0,22
Khopseksky massif													
L -69	Clinopyroxenite	0,14	0,40	0,40	0,14	0,05	0,33	0,04	0,03	0,18	0,03	1,74	0,52
L -70	-"-	0,96	2,80	2,00	0,79	0,29	0,95	0,14	0,08	0,43	0,05	8,49	1,52
L -78	-"-	0,24	0,60	0,60	0,15	0,11	0,29	0,04	0,04	0,27	0,05	2,39	0,61
L -84	-"-	0,24	0,40	0,20	0,09	0,04	0,33	0,02	0,02	0,09	0,02	1,45	1,74
Irbiteysky massif													
L -66	Olivine gabbro	0,98	2,50	1,90	0,67	0,32	0,91	0,16	0,09	0,58	0,09	8,20	1,14
L -68	-"-	0,74	1,90	1,80	0,64	0,32	1,30	0,15	0,10	0,54	0,09	7,58	0,92
Maysky massif													
2530	Olivine gabbro	3,80	9,00	5,20	1,48	0,70	1,60	0,26	0,12	0,74	0,11	23,01	3,47
2150	Gabbro-norite	5,00	10,60	6,60	1,85	0,84	1,60	0,29	0,18	1,14	0,17	28,27	2,96

Note. Analyses were done in Analytical Center of IGM SB RAS, LA ICP-MS method (analyst M.S. Mel'gunov.

Table 2. Platinum, palladium, and rhodium composition of rocks from some massifs, Central Tuva, (Russia) (ppb)

Sample's number	Pt	Pd	Rh	(Pt+Pd+Rh)	Pd/Pt
Mazhaliksky massifs					
L-106	29	35	4	68	1,21
L-106a	25	36	4	65	1,44
L-101a	15	38	5	48	1,87
Mazh-35-5	19	21	5	43	1,11
L-104A	9	30	4	43	3,33
Mazh -30-4	23	29	5	55	1,26
Karaossky massifs					
M-35-97	15	19	5	39	1,27
M-26-98	16	18	5	39	1,13
M -36-98	12	12	5	29	1,00
M -25-97	20	29	5	53	1,45
Kalbagdagsky massif					
Kal-3-3	42	40	4	86	0,95
Kal-6-3	35	40	4	79	1,14
Kal-8-1	41	52	5	98	1,27
Kal-9-2	30	44	4	78	1,47
Kal-5-2	28	54	7	89	1,93
Kal-3-2	29	61	8	98	2,10
Kal-10-3	25	80	7	112	3,20
Kal-12-4	17	23	5	44	1,35
L-31	9	28	4	41	3,11
Karashatsky massif					
L-50	48	59	6	113	1,23
L-50a	39	80	6	125	2,05
L-37	8	8	3	19	1,00
L-48	78	97	7	182	1,24
L-36	10	56	6	72	5,60
L-44	20	36	4	60	1,80
L-45	28	43	5	76	1,54
Khopseksky massif					
L-69	13	31	4	48	2,38
L-70	17	48	5	70	2,82
L-78	23	32	4	59	1,39
L-84	40	90	6	136	2,25
L-84a	36	70	6	112	2,16
Irbiteysky massif					
L-66	15	66	4	85	4,40
L-68	15	44	4	63	2,93
Maysky massif					
2530	58	17	7	82	0,29
2150	36	10	5	51	0,28
Lower detection limits	10	5	5	–	–

Note. Sample's number and rock's names are correspond Table 10.1. Analyses were done in Analytical Center of IGM SB RAS, atom-absorption method (analyst V.G. Tsimbalist).

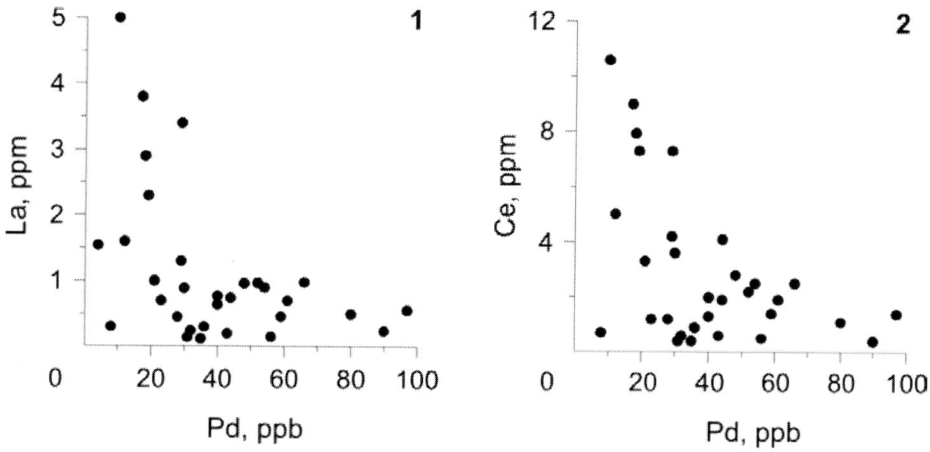

Figure 1. The graphs of correlation between La и Pd compositions (1), and Ce и Pd compositions (2) for rocks from some mafic-ultramafic massifs of Central Tuva (Russia) (data Tables 1 and 2).

Table 3. Modal mineral composition of rocks from some massifs, Central and East Tuva (Russia)

Sample's number	Rocks	Composition, %
Mazhaliksky massif		
30-1	Gabbro leucocratic with hornblende	Cpx (17), Pl (80), Horb (3)
30-6	Olivine gabbro	Ol (11), Cpx (45), Pl (50), Mt (4)
35-4	Wherlite Pl-bearing with Horb	Ol (50), Cpx (40), Pl (4), Horb (6)
Karaossky massif		
M-27-98	Websterite Pl-bearing	Opx (36), Cpx (60), Pl (4)
M-29-98	Gabbro-norite	Opx (10), Cpx (40), Pl (50)
M-30-98	Wherlite with Opx	Ol (85), Opx (3), Cpx (12)
Kalbagdagsky massif		
9-3	Gabbro olivine-bearing with Horb	Ol (20), Cpx (45), Pl (34), Horb (1)
1-2	Gabbro-norite	Cpx (35), Opx (15), Pl (50)
2-5	Clinopyroxenite olivine-bearing with Horb	Ol (8), Cpx (80),Pl (4), Horb (8)
Maysky massif		
2150	Gabbro-norite	Cpx (30), Opx (15), Pl (55)
2530	Gabbro olivine-bearing, leucocratic with Horb	Ol (20), Cpx (8), Pl (70), Horb (2)
2530-1	Gabbro olivine-bearing, mesocratic with Opx	Ol (40), Opx (4), Cpx (10), Pl (45), Mt (1)
Bulkinsky massif		
Bl-4	Gabbro olivine-bearing, leucocratic with Horb	Ol (10), Cpx (15), Pl (70), Horb (4), Mt (1)
Bl-7	Troctolite leucocratic with Cpx, Opx, and Horb	Ol (20), Opx (3), Cpx (1), Pl (70), Horb (5), Mt (1)
Bl-17	Gabbro-anorthosite with Ol, Cpx, Opx, and Horb	Ol (2), Opx (2), Cpx (2), Pl (80), Horb (10), Mt (1)

Note. Ol – olivine, Opx – orthopyroxene, Cpx – clinopyroxene, Pl – plagioclase, Horb – hornblende, Mt – magnetite.

long axis oriented to the submeridional direction. The massif lies among terrigenous Upper Cambrian and Ordovician deposits and is composed of gabbro-anorthosites, anorthosites, gabbros, gabbronorites and troctolites, among which stripe-like isolations of olivinites, olivine pyroxenites and hornblendites occur. Quantitative-mineral composition of rocks from this and from other massifs, in which REE contents were determined by ICP-MS method, is shown in the Table 3.

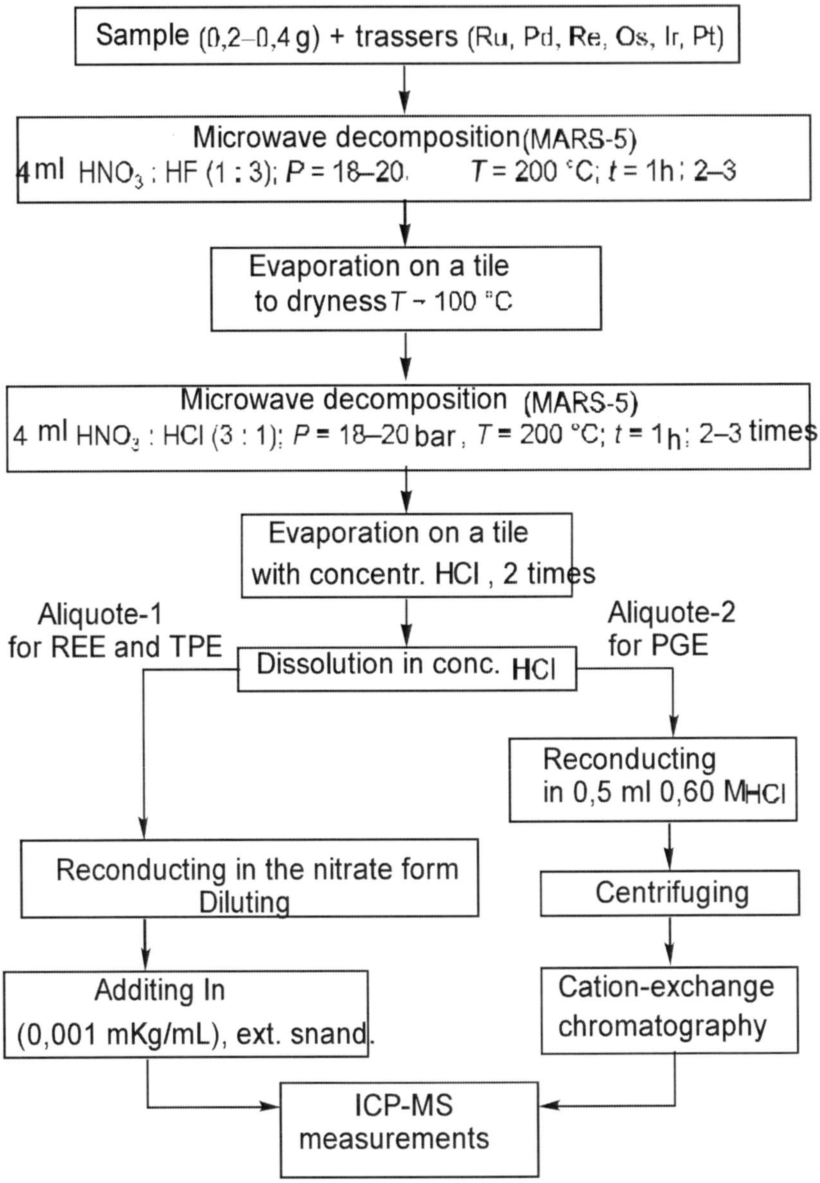

Figure 2. The scheme of method of chemical preparation of rock samples for REE, rare elements and platinum element group (PGE) analysis (ICP-MS) [Palessky, 2008]. DME – difficult-melting elements (Zr, Nb, Ta, Hf).

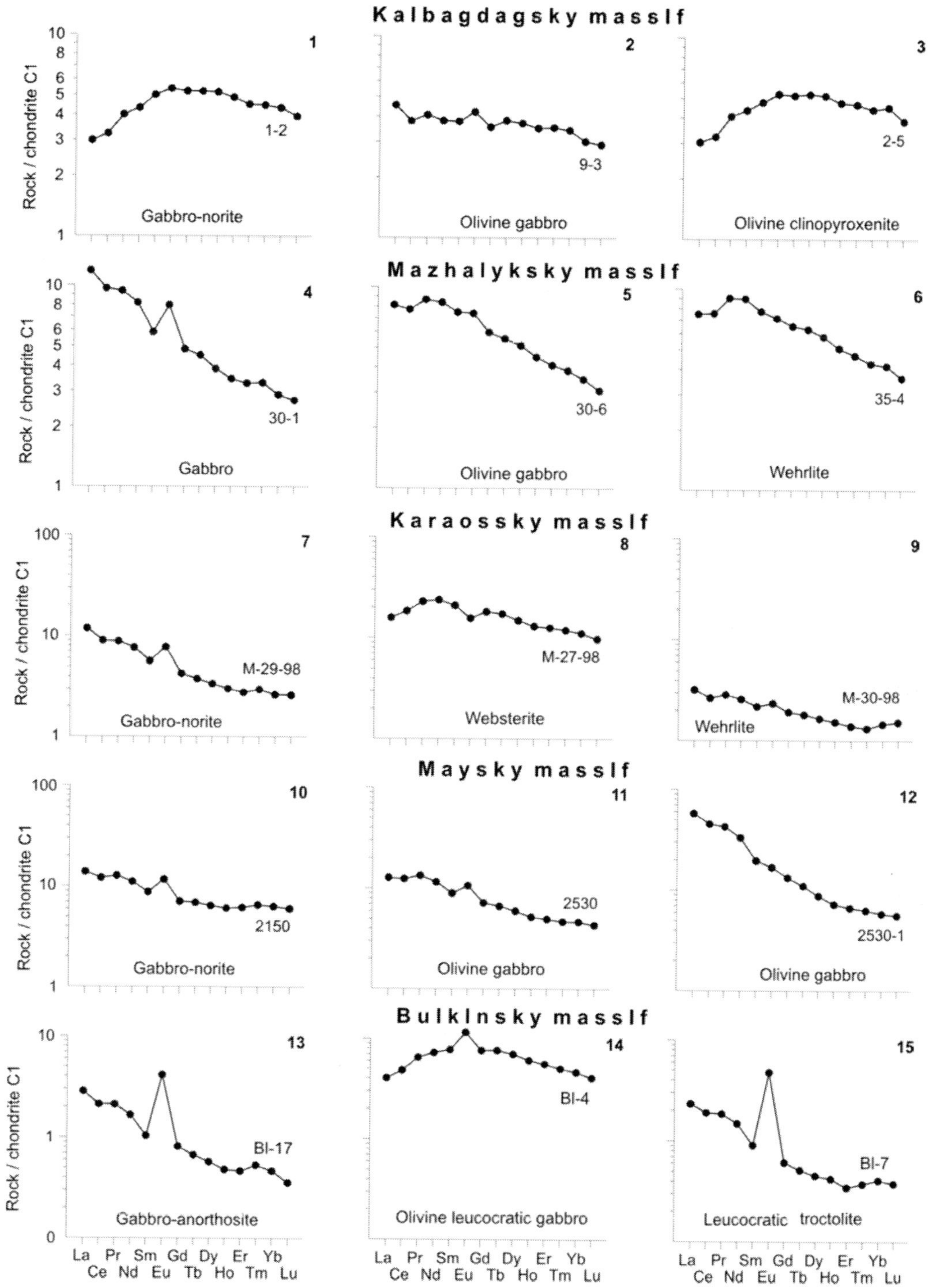

Figure 3. Chondrite-normalized REE patterns for rocks from Kalbagdagsky, Mazhalyksky, Karaossky, Maysky, and Bulkinsky mafic-ultramafic massifs (Tuva, Russia) (data Table 4). Here and further the normalizing of REE contents were carry out using their contents in the chondrite C1 [Anders, Grevesse, 1989].

Table 4. Rare earth element (ppm), and platinum element group (ppb) composition of rocks from some massifs in Central and East Tuva (Russia)

	Massifs						
	Kalbagdagsky			Mazhalyksky			Karaossky
Element	1-2	9-3	2-5	30-1	30-6	35-4	M-29-98
	Gabbro-norite	*Olivine gabbro*	*Clinopyroxenite olivine-bearing*	*Gabbro*	*Olivine gabbro*	*Wherlite*	*Gabbro-norite*
Pd	7,60	8,77	5,69	3,28	3,74	8,51	N.d.
Pt	25,4	25,7	9,27	18,4	7,52	2,65	8,39
Ru	4,23	2,09	5,03	1,74	0,75	3,76	3,95
Ir	0,097	0,046	0,067	0,29	0,029	N.d.	0,19
Total	37,3	37,0	20,1	23,7	12,0	14,9	12,5
Pt/Ir	262	558	138	63	252	N.d.	44
Pd/Ir	78,4	191	84,9	7,86	129	»	N.d.
La	0,729	1,11	0,742	2,88	1,99	1,84	2,89
Ce	2,05	2,42	2,06	6,15	4,94	4,83	5,71
Pr	N.d.	N.d.	N.d.	N.d.	N.d.	N.d.	N.d.
Nd	2,05	1,81	2,07	3,89	3,98	4,24	3,62
Sm	0,772	0,581	0,737	0,900	1,16	1,19	0,872
Eu	0,312	0,244	0,305	0,463	0,430	0,415	0,448
Gd	1,07	0,724	1,06	0,984	1,21	1,34	0,872
Tb	0,195	0,143	0,196	0,168	0,207	0,236	0,143
Dy	1,32	0,941	1,31	0,979	1,30	1,47	0,863
Ho	0,276	0,198	0,270	0,195	0,254	0,288	0,172
Er	0,748	0,586	0,776	0,542	0,679	0,778	0,462
Tm	0,114	0,088	0,113	0,084	0,099	0,109	0,076
Yb	0,715	0,498	0,747	0,474	0,576	0,686	0,439
Lu	0,100	0,074	0,098	0,068	0,078	0,092	0,067
(La/Yb)*n*	0,69	1,51	0,67	4,11	2,34	1,81	4,45
(Eu/Eu*)*n*	1,05	1,15	1,05	1,50	1,10	1,00	1,55

Judging by the results of the analysis, in the rocks of the studied massifs the total REE contents ranged from 3.5 to 75 ppm (Table 4). And the maximum total REE contents were determined in samples from Maysky massif, the lower ones - in the rocks from Mazhalyksky, Kalbagdagsky, Karaossky and Bulkinsky massifs. The rocks of Maysky and Bulkinsky massifs are characterized by high LREE content and the resulting relatively high values of (La/Yb)*n* parameter (Figure 3). The REE distribution patterns of clinopyroxene and olivine gabbro-norite of Kalbagdagsky massif have similar configuration, which indicates their depletion by LREE and some enrichment with MREE. The REE distribution pattern of olivine gabbro of the massif also indicates its enrichment with LREE. A distinctive feature of the rocks from Kalbagdagsky massif is the lack of positive Eu anomalies in their patterns, which indicates that during the crystallization of their parent melt, no significant fractionation of plagioclase crystals occurred (Figure 3, 1-3).

The REE patterns of gabbro, olivine gabbro and wehrlite of Mazhalyksky massif that have steep negative slopes, are complicated by positive Eu anomalies due to fractionation of plagioclase (Figures 3, 4-6). In the rocks of Karaossky massif the lowest REE contents are observed in wehrlite, slightly higher in gabbro-norite, and much higher in websterite. The REE distribution patterns in all these rocks have a common negative slope (Figures 3, 7-9). It was already mentioned, that rocks of Maysky massif differ from samples of other massifs by a generally high REE content and especially LREE content. Similar configuration of REE patterns of gabbro-norite and olivine gabbro from this massif shows weak positive Eu anomalies, while the pattern of olivine gabbro from the massif that has a steep negative slope doesn't show such anomaly (Figures 3, 10-12). We can assume that the anomalous enrichment with LREE of the rocks from Maysky massif is due to the presence of non-structural impurity of these elements in them, which was introduced by fluids from the melts that formed the granitoids pluton located nearby.

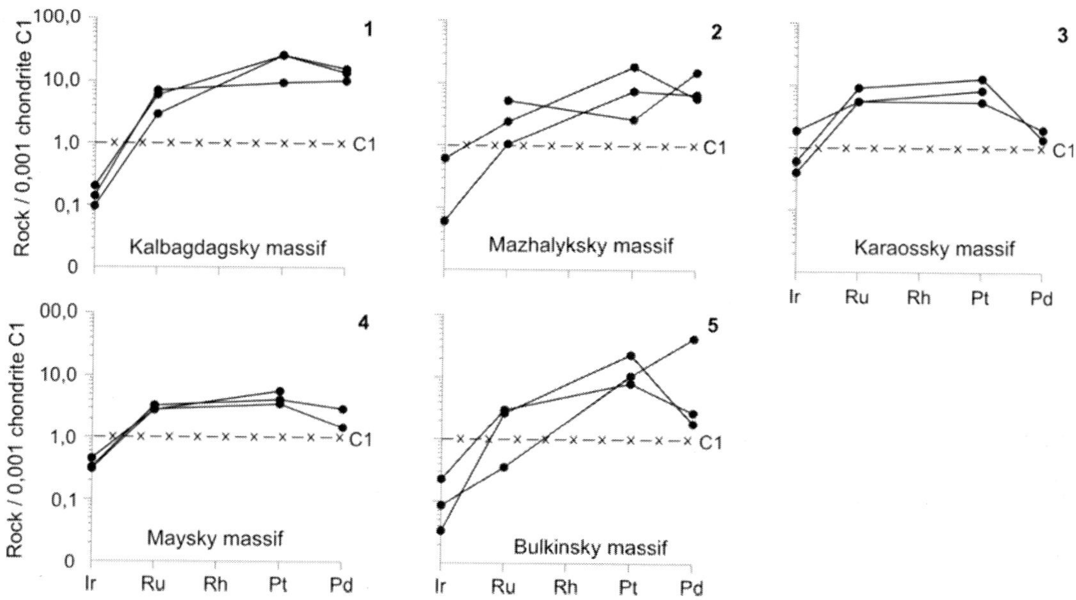

Figure 4. Chondrite-normalized PGE patterns for rocks from some mafic-ultramafic massifs in Central and East Tuva (Russia) (data Table 4). Here and further the normalizing of REE contents were carry out using their contents in the chondrite C1 [Anders, Grevesse, 1989].

The rocks of Bulkinsky massif are also enriched with LREE. Their distribution patterns of gabbro-anorthosite and leucocratic troctolite are complicated by positive Eu anomalies of moderate intensity. Olivine leucocratic gabbro that is significantly enriched with all REE, has a pattern complicated by especially intense positive Eu anomaly (Figures 3, 13-15).

The determination of Pd, Pt, Ru and Ir in rocks of the massifs located in the territory of Tuva performed by ICP-MS method showed that the total PGE content in them ranges from 6.0 to 38 ppb. Contents of certain elements vary in the following ranges (ppb): Pt (2.65-26.7); Pd (0.82-24.0); Ru (0.25-6.56); Ir (0.02 – 0.90). In most samples the content of Pt is higher than the content of Pd, with the exception of wehrlites of Mazhalykskу massif and gabbro – anorthosite from Bulkinsky massif. In the rocks of Kalbagdagsky and Mazhalykskу massifs the Pd content is higher than the content of Ru, but in the rocks of Karaossky, Maysky and Bulkinsky massifs, with the exception of Bl-17 sample, the content of Ru is higher that of Pd content. In addition, the rocks of Karaossky massif are characterized by relatively high Ir content. The mentioned features of PGE geochemistry in rocks of the studied massifs are reflected in their chondrite-normalized patterns (Figure 4). Almost all of these patterns are arched upwards, which indicates a slight enrichment of these rocks with platinum and ruthenium relative to palladium and iridium.

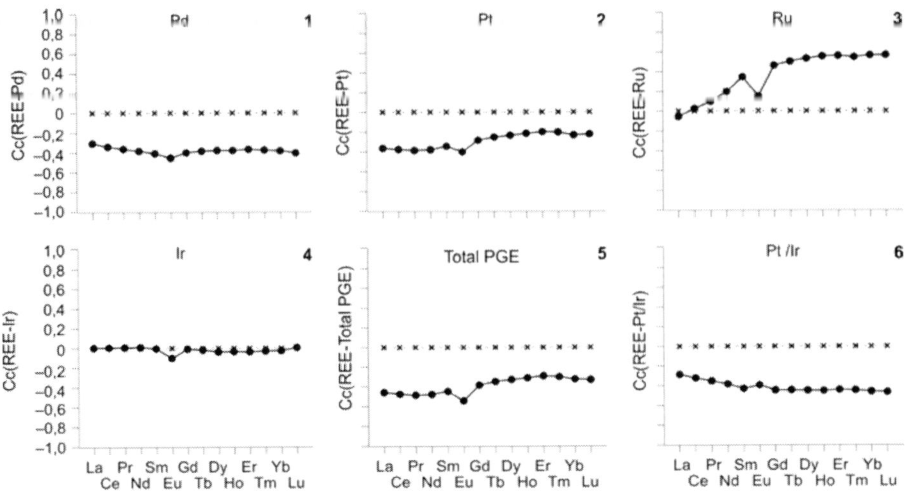

Figure 5. Graph of correlation coefficients (Cc) between REE and PGE composition for rocks from some mafic-ultramafic massifs of Central and East Tuva (Russia) (data Table 4).

The Pt/Ir parameter, as it was already mentioned, can be used as a geochemical indicator of estimating the degree of PGE fractionation during the formation of rocks containing them. In the rocks of mafic-ultramafic massifs of Tuva the values of this parameter vary in a wide range - from 6 to 1450. By using these values the massifs can be divided into two groups. The first group includes those massifs, the rocks of which have higher values of this parameter: Kalbagdagsky (138-558), Bulkinsky (74-1450) and Mazhalykský (63-252). The second group consists of massifs of rocks that are characterized by low values of this parameter: Maysky (23-26) and Karaossky (6-45).

Data on the distribution of REE and PGE in rocks of the massifs of Tuva obtained by ICP-MS method allow reconsidering the links between these two groups of chemical elements. Using the full sample of analyses of these rocks we have calculated the coefficients of pair correlation between the contents of certain REE and PGE, which are presented in the form of the following graphs (Figure 5). They imply that in almost all cases the correlation coefficients between Pd and Pt, on the one hand, and certain REE, on the other hand, have negative values that vary in the range from -0.20 to -0.40. However, all these coefficients are statistically insignificant since being less than the critical value of this sample (0.48) (Figures 5, 1). Sub horizontal position of the graphs of changing of correlation coefficients indicates that the negative correlation between LREE and HREE, on the one hand, and Pd, on the other hand, have approximately the same intensity. Slight positive slope of the curve of changing of correlation coefficients between REE and Pt indicate that the latter has a slightly stronger negative correlation with LREE compared to HREE (Figures 5, 2). The type of the graph of changing of coefficients for Ru indicates that this element has a positive relation with all REE, except for La and Ce (Figures 5, 3). A relatively small number of available tests did not allow determining the nature of relationship between REE and Ir (Figures 5, 4). In contrast to the statistically insignificant negative correlation coefficients between total PGE contents and most of the REE, the values for Eu (-0.54) are statistically significant (Figures 5, 5). We can also assume a negative correlation between HREE and Pt/Ir parameter (Figures 5, 6). The possibility of an inverse relation between REE and PGE contents in these rocks is confirmed by the location of figurative points in Figure 6.

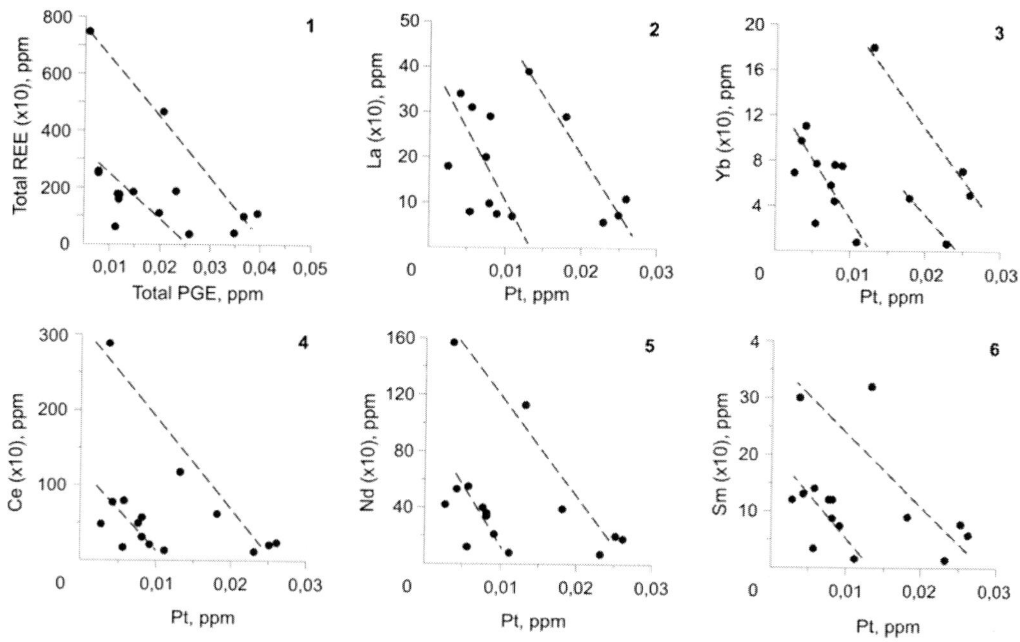

Figure 6. Graphs of correlation between REE and total PGE and Pt composition for rocks from some mafic-ultramafic massifs in Central and East Tuva (Russia) (data Table 4). When constructing graphs REE content increased by 10 times. Here and below the dotted lines - estimated trends.

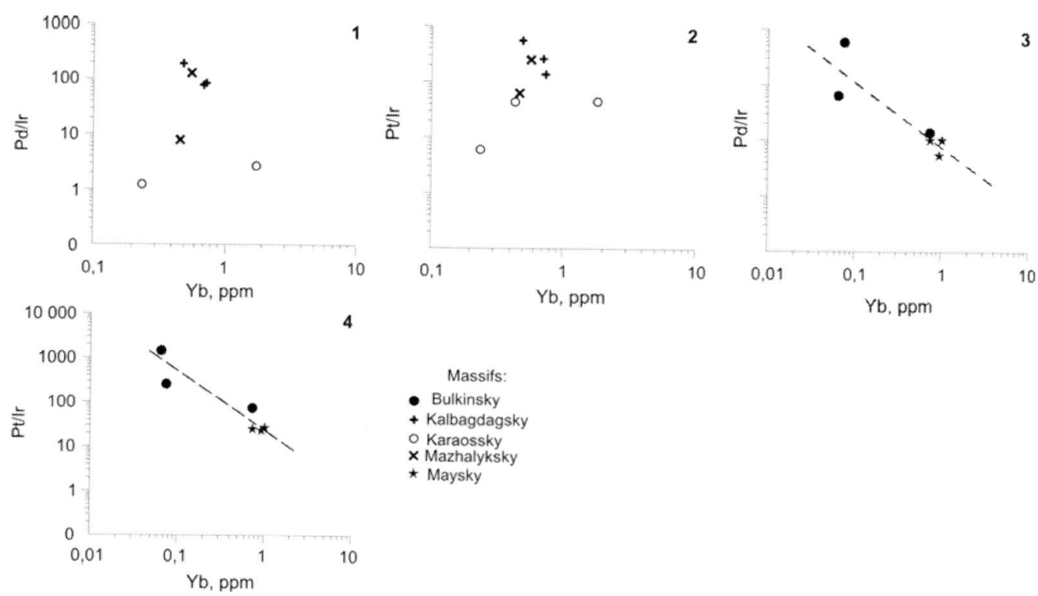

Figure 7. Graphs of correlation between Yb composition and Pd/Ir parameter values (1 and 3) and Pt / Ir parameter values (2 and 4) for rocks from Kalbakdagsky, Mazhalyksky, Karaossky, Maysky and Bulkinsky massifs (Tuva, Russia) (data Table 4.).

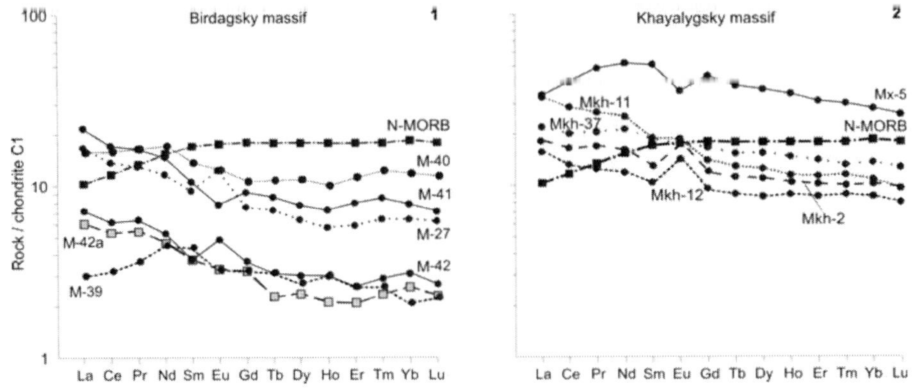

Figure 8. Chondrite-normalized REE patterns for rocks from Birdagsky (1) and Khayalygsky (2) massifs (Tuva, Russia) (data Table 5).

These data allow us to refer to the ratio of two indicator parameters - Pd/Ir and Pt/Ir. Thus, in the rocks of Kalbagdagsky massif both of these parameters have higher values if compared with rocks from Karaossky and Maysky massifs. Leucocratic troctolite from Bulkinsky massif has the highest value of Pt/Ir parameter (see Table 10.4). Judging by the diagrams in the coordinates (Pd / Ir-Yb) and (Pt / Ir-Yb), there can be assumed a definite relationship between these two indicators of fractionation of PGE and Yb (Figure 7).

Next, we consider the possible relationship between REE and PGE in rocks of two more massifs - Birdagsky and Khayalygsky. The data obtained indicate that the rocks of Birdagsky massif have lower total REE contents in comparison with the rocks of Khayalygsky massif (Table 5). The level of HREE and MREE accumulation in the rocks of both massifs is lower than in N-MORB basalts, while the level of LREE accumulation is higher, on the contrary. By the nature of the patterns of chondrite-normalized REE contents we can conclude that LREE contents in these rocks are slightly higher than HREE contents and that in some samples there is a slight Eu excess (Figure 8). By the content of REE the rocks of Birdagsky massif are divided into two types. The first is represented by hornblende-bearing gabbro containing heightened concentrations of these impurities in comparison with the rocks of the second type – plagioclase-bearing lherzolites and wehrlites. The rocks of Khayalygsky massifs can also be divided into two geochemical types: 1) hornblende-bearing gabbro with lower REE content, and 2) hornblendites that have heightened contents of them. In general, almost all samples of these two massifs have higher REE concentrations in comparison with rocks that form the massifs located in central and eastern regions of Tuva.

According to data obtained by atomic absorption method, in rocks of Birdagsky and Khayalygsky massifs the total Pt, Pd and Rh contents range from 20 to 74 ppb. At the same time, in some samples the content of Pt is higher than Pd content, while in the others – the opposite situation. According to the analysis by ICP-MS method in the rocks of both massifs the total Os, Ir, Ru, Pt and Pd contents ranged from 0.6 to 21 ppb (Table 5). In plagioclase-bearing lherzolites and wehrlites of Birdagsky massif, as well as in amphibole gabbro of Khayalygsky massif the level of PGE accumulation is generally higher than in the rest of samples. Total Pt and Pd contents in all rocks of both massifs exceed the total contents of Os,

Table 5. Rare earth element, rare element (ppm), platinum group element, and rhenium (ppb) composition of rocks from Birdagsky and Khayalygsky massifs (South-West Tuva, Russia) [Oydup et al., 2007]

	Massifs										
	Birdagsky						Khayalygsky				
Element	M-42	M -42a	M -39	M -40	M -27	M -41	M-kh-5	Mkh -37	Mkh -11	Mkh -12	Mkh -2
	1	2	3	4	5	6	7	8	9	10	11
La	1,73	1,47	0,72	3,92	4,20	5,48	8,19	5,31	8,14	3,92	4,48
Ce	3,92	3,40	1,98	10,34	9,02	11,22	25,55	12,75	18,59	8,49	10,60
Pr	0,61	0,53	0,34	1,62	1,28	1,63	4,64	1,95	2,62	1,22	1,66
Nd	2,49	2,23	2,09	8,29	5,69	7,17	24,38	10,03	12,17	5,72	7,67
Sm	0,57	0,58	0,66	2,14	1,49	1,69	7,80	2,87	2,96	1,62	2,00
Eu	0,28	0,19	0,19	0,74	0,74	0,46	2,05	1,01	1,10	0,85	1,04
Gd	0,73	0,65	0,64	2,22	1,59	1,92	8,80	3,38	2,91	1,95	2,44
Tb	0,12	0,08	0,11	0,41	0,28	0,33	1,41	0,57	0,49	0,34	0,42
Dy	0,77	0,60	0,67	2,83	1,67	2,01	9,15	3,88	3,22	2,21	2,82
Ho	0,17	0,12	0,17	0,58	0,33	0,42	1,93	0,83	0,66	0,51	0,60
Er	0,43	0,34	0,42	1,91	1,00	1,34	5,12	2,31	1,90	1,45	1,71
Tm	0,07	0,06	0,07	0,32	0,17	0,22	0,76	0,33	0,30	0,23	0,26
Yb	0,51	0,43	0,34	2,00	1,08	1,34	4,62	2,23	1,82	1,45	1,71
Lu	0,07	0,06	0,06	0,30	0,16	0,19	0,66	0,32	0,25	0,20	0,25
Total REE	12,5	10,7	8,46	37,6	28,7	35,4	105	47,8	57,12	30,16	37,65
$(La/Yb)_n$	57,8	58,4	35,6	35,0	55,4	60,2	39,0	39,5	74,82	41,45	42,44
$(Eu/Eu^*)_n$	1,34	0,94	0,87	1,03	1,46	0,79	0,75	0,99	1,13	1,46	1,44
Rb	4,57	3,82	1,32	58	27	5,93	6,27	37	71	34	75
Sr	62	67	35	67	323	201	99	312	379	324	438
Y	5,13	4,15	4,51	21	11,9	13,7	61	25	22	14,6	18,8
Zr	21	14,8	4,70	20	40	30	90	74	86	37	43
Nb	1,65	1,02	0,13	1,61	2,48	1,00	8,00	7,38	5,04	2,91	3,68
Cs	7,91	6,74	0,45	3,71	4,17	2,39	0,60	1,62	2,66	2,13	2,69
Ba 18,80		31	11,08	259	279	86	89	230	386	270	1219
Hf	0,50	0,33	0,16	0,65	0,89	0,82	2,95	1,94	2,02	1,00	1,25
Ta	0,19	0,28	0,05	0,091	0,18	0,18	0,46	0,45	0,36	0,19	0,37
Th	0,46	0,18	0,09	0,27	0,89	0,63	0,90	0,71	1,68	0,37	0,37
U	0,17	0,084	0,082	0,081	0,24	0,16	0,33	0,32	0,48	0,17	0,083
Sr/Ba	3,30	2,13	3,20	0,26	1,16	2,33	1,12	1,35	0,98	1,20	0,36
Sr/Rb	13,6	17,5	26,9	1,15	3,56	33,9	15,8	8,47	5,31	9,59	5,82
Zr/Hf	42,9	44,2	28,7	30,8	44,2	36,8	30,6	38,3	42,7	36,6	34,7
Zr/Y	4,17	3,56	1,04	0,97	3,33	2,20	1,49	2,99	3,96	2,50	2,31
Ta/Nb	0,11	0,28	0,40	0,06	0,07	0,18	0,06	0,06	0,07	0,06	0,05
U/Th	0,36	0,46	0,91	0,30	0,27	0,26	0,36	0,46	0,29	0,46	0,23
Os	0,08	0,03	0,62	0,02	0,02	N.d.	0,03	0,43	0,17	0,51	0,07
Ir	0,06	0,07	0,16	0,04	0,02	0,01	0,02	0,05	0,10	0,09	0,02
Ru	0,62	0,60	0,55	0,51	0,12	0,47	0,09	0,91	1,78	0,55	0,58
Pt	5,28	3,33	19,0	0,93	0,09	0,60	0,10	0,91	3,77	1,12	0,34
Pd	4,96	5,34	0,70	0,80	0,36	1,04	0,32	1,54	4,43	2,68	0,98
Total PGE	11,00	9,37	21,03	2,30	0,61	2,12	0,56	3,84	10,25	4,95	1,99
Re	0,36	0,36	0,16	0,08	0,17	0,26	0,21	0,12	0,48	0,41	0,19
Pd/Pt	0,94	1,60	0,04	0,86	4,00	1,73	3,20	1,69	1,18	2,39	2,88
Pd/Ir	82,7	76,3	4,4	20,0	18,0	104	16,0	30,8	44,3	29,8	49,0
Pt/Ir	88,0	47,6	119	23,3	4,5	60,0	5,00	18,2	37,7	12,4	17,0
Ru/Ir	10,3	8,57	3,44	12,8	6,00	47,0	4,50	18,2	17,8	6,11	29,0

Note: 1, 2 – lherzolites Pl-bearing, serpentinized; 3 – olivine clinopyroxenites (wherlite); 4 – hornblendite Pl-bearing; 5 – hornblende gabbro; 6 – olivine gabbro-norite with hornblende, melanocratic; 7 – hornblende Pl-bearing, coarse-grained; 8 – hornblende gabbro, melanocratic; 9 – hornblende gabbro, coarse-grained; 10 – hornblende gabbro; 11 – hornblende gabbro with phlogopite. Analyses were done in Analytical Center of IGM SB RAS, ICP-MS method (analysts S.V. Palessky, I.V. Nikolaeva, O.A. Koz'menko). Detection limits of PGE (ppb): Os – 0,07; Ir – 0,002; Ru – 0,005; Pt – 0,07; Pd – 0,09. N.d. – no data.

Ir and Ru. In all rocks of Khayalygsky massif and some samples from Birdagsky massif the content of Pd is higher than Pt content, in the rest of the rocks from Birdagsky massif the reaction between these elements is inverse. Plagioclase-bearing lherzolites of Birdagsky massif and also some hornblende-bearing gabbro from Khayalygsky massif are characterized by slightly heightened Re concentrations.

The chondrite-normalized patterns of PGE and Re contents in the rocks of Birdagsky and Khayalygsky massifs are similar in configuration and have a common positive slope (Figure 9). At the same time some of the patterns are complicated by minima for Ir and Pt, and minima for Ir in the patterns of rocks from Khayalygsky massifs have a slightly greater intensity than in the other cases. The values of Pd/Pt parameter in rocks from both of these massifs vary in a narrow range compared with the values of Pd/Ir and Pt/Ir parameters. In the rocks of Birdagsky massif the relationship between the values of Pd/Ir and Pt/Ir parameters, on the one hand, and Yb contents, on the other hand, has not been ascertained, while in the rocks of Khayalygsky massif an inverse relationship between these characteristics should not be excluded (Figure 10).

Based on data on the relations between REE and PGE in rocks, the mafic-ultramafic massifs of Tuva can be divided into three types: 1) massifs of rocks that are relatively enriched with REE, but significantly depleted by PGE; 2) massifs of rocks that are relatively depleted by REE and relatively enriched with PGE; 3) massifs of rocks that are characterized by low concentrations of both REE and PGE. An inverse relationship between the contents of REE and PGE occurs primarily in the rocks of the first two types of massifs. The above material on geochemistry of REE and PGE in rocks of mafic-ultramafic massifs located in the territory of Tuva, with some exceptions, generally does not contradict the assumption of the existence of a sort of antagonism between these two groups of elements in different products of mafic-ultramafic magmatism.

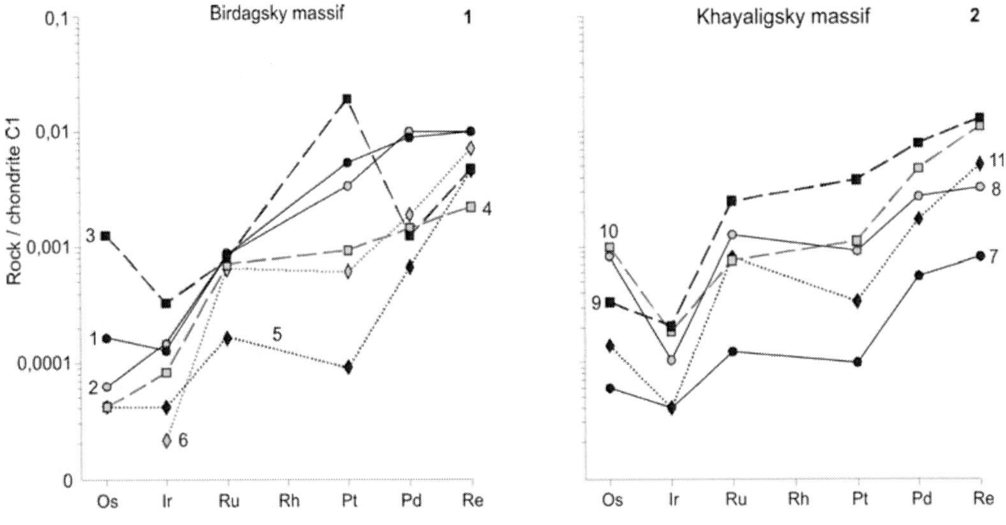

Figure 9. Chondrite-normalized PGE patterns for rocks from Birdagsky (1) and Hayalygsky (2) massifs (Tuva, Russia) (data Table 10.5). Numbers of graphs correspond to the ordinal numbers of samples in Table 5.

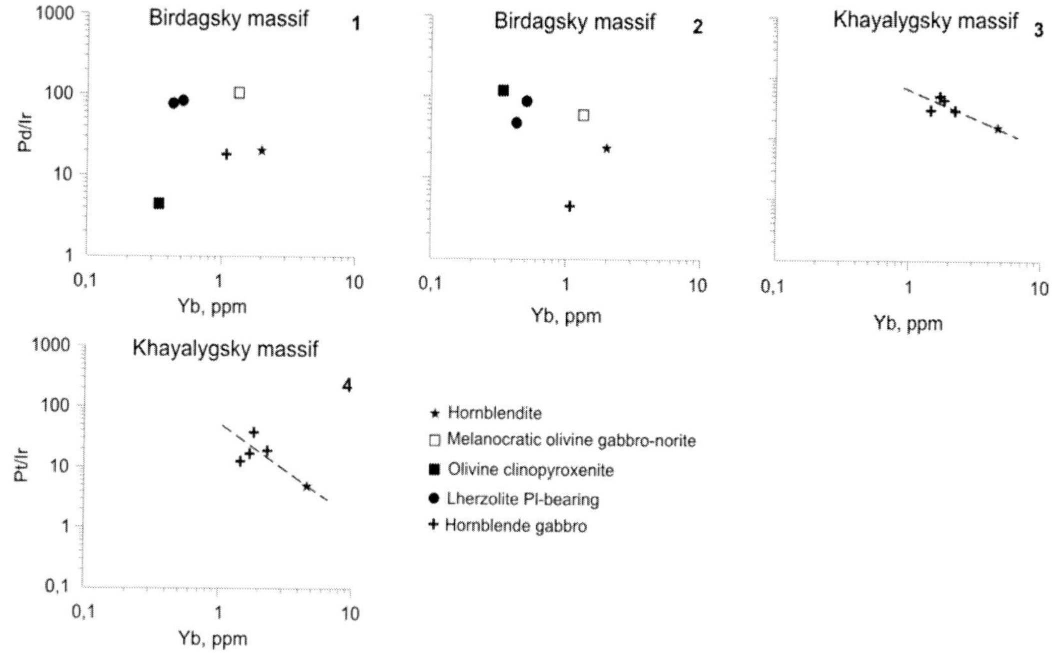

Figure 10. Graphs of correlation between Yb composition and Pd/Ir parameter values (1 and 3) and Pt / Ir parameter values (2 and 4) for rocks from Birdagsky and Khayalygsky (Tuva, Russia) (data Table 5.).

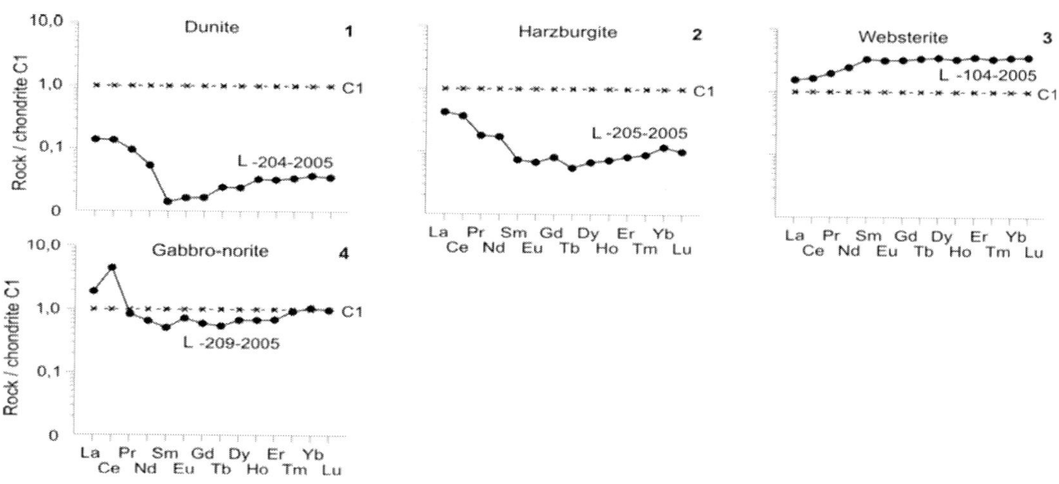

Figure 11. Chondrite-normalized REE patterns for rocks from Naransky massif (Western Mongolia) (data Table 6).

Naransky massif (Western Mongolia) located 15 km. south-east of the Gobi Altai city is one of the largest mafic-ultramafic massifs in the territory of Western Mongolia. It is a part of Dzabkhan ophiolite belt that traces the zone of abyssal fracture. The massif is extended to the west-northwest direction at 38 km. at a width of 0.5-10 km. Its structural position, internal structure, petrographic, mineralogical, petrochemical, geochemical and metallogenic features were discussed in many publications [Pinus et al., 1976, 1984; Lesnov, 1979, 1981, 1982;

Lesnov & Simonov, 1989; Dmitrenko et al., 1991] Below, we discuss the previously obtained data on the distribution of REE and PGE in rocks of the massif [Lesnov & Tsimbalist, 1983; Lesnov, 1993], as well as new materials related to this issue. Naransky massif is a complex plutonic body, the structure of which contains three main structural elements [Lesnov, 1982]: 1) the large steeply dipping ultramafic restite protrusion, and 2) intruding through this protrusion along its southern tectonic contact, much inferior to it in size and also steeply dipping gabbro intrusion of elongated form, and 3) the contact-reaction zone localized along the boundaries of ultramafic protrusion and gabbro intrusion, which is composed of hybrid rocks of ultramafic and mafic composition. To the south of the massif, among the host strata there are several small tectonic blocks of ultramafic restites and gabbro stocks intruding through them and the host strata. The host a stratum is represented by metamorphosed terrigenic-volcanic and carbonate rocks of the Vendian-Cambrian age.

Naransky massif and its host strata are transgressive overlain by horizons of vulcanite of average composition and conglomerates of Devonian age. The later formation of gabbro intrusion in relation to ultramafic protrusion is confirmed by the contact-reaction zone located along their borders, xenoliths of ultramafic restites in the composition of gabbro intrusion found in several places, as well as their veined bodies of gabbro and pyroxenites injecting them. Ultramafic restites are represented by harzburgites and by less common lherzolites and dunites, stripe-like bodies of which alternate with harzburgites. All these rocks are serpentinized and dinamometamorphosed on different levels, especially when close to the faults. Gabbro intrusion and its stock-like satellites are composed of gabbronorites and gabbro, as well as olivine-bearing varieties of them. The structure of the contact-reaction zone involves websterites that alternate with lenticular and stripe-like isolations of wehrlites and olivine gabbro of variable quantitative-mineral composition.

Among the ultramafic restites many diluvial-eluvial and as well as some indigenous manifestations of ore chromitites there were found. In websterites of the contact-reaction zone there was scattered sulphide mineralization identified, which contains microparticles of copper-bearing minerals of Pd and Pt. In many schlich samples from the alluvium and proluvium of temporary channels draining the massif there were grains of PGE minerals of various composition, as well as Au were found [Lesnov, 1982; Dmitrenko et al., 1991]. The distribution of REE, PGE and Au in rocks of Naransky massif has been studied using atomic absorption method, as well as RNAA and ICP-MS methods (Table 6, 7). According to the available analyses, the lowest total REE contents are characteristic of dunites (0.18 ppm). Note that approximately the same average total REE content (0.20 ppm) was determined in dunites from the Twin Sisters massif (USA) [Frey et al., 1971], as well as in the international geochemical standard of DTS-1dunite (0.17 ppm) [Gladney et al., 1991].

The higher total REE content is defined in harzburgites from this massif – 0.53 ppm. To add more, an approximately same level of total REE content was determined in harzburgites of Ronda massif (Spain) - from 0.34 to 1.08 ppm [Van der Wal, Bodinier, 1996]. In websterites of Naransky massif the total REE contents are much higher (6.66 ppm) than in dunites and harzburgites, but lower than in websterites of Balmuccia massif (Italy) [Rivalenti et al., 1995].

Table 6. Rare earth element, rare element (ppm), platinum group element, and rhenium (ppb) composition of rocks from Naransky massif (West Mongolia)

Element	L-204	L-205	L-104	L-209	266	269	278	Primitive mantle**
	Methods							
	ICP-MS				RNAA			
	Dunite	*Harzburgite*	*Websterite*	*Gabbro-norite*	*Websterite*	*Gabbro-norites*		
La	0,034*	0,10	0,38	0,47	0,250	N.d.	0,48	0,687
Ce	0,087*	0,23	1,04	2,9	0,330	0,42	0,6	1,775
Pr	0,008*	0,016*	0,18	0,075	N.d.	N.d.	N.d.	0,276
Nd	0,026*	0,082	1,16	0,31	0,230	0,28	0,3	1,354
Sm	0,002*	0,011*	0,51	0,079	0,064	0,1	0,13	0,444
Eu	0,0009*	0,004*	0,18	0,042	0,027	0,1	0,1	0,168
Gd	0,003*	0,017	0,66	0,12	0,120	0,24	0,27	0,596
Tb	0,0009*	0,002*	0,13	0,021	0,023	0,042	0,05	0,108
Dy	0,006*	0,017	0,89	0,17	N.d.	N.d.	N.d.	0,737
Ho	0,002*	0,004*	0,19	0,038	»	»	»	0,164
Er	0,005*	0,014	0,59	0,11	»	»	»	0,480
Tm	0,0009*	0,002*	0,086	0,024	0,027	0,039	0,046	0,074
Yb	0,006*	0,020	0,58	0,18	0,150	0,25	0,29	0,493
Lu	0,0009*	0,003*	0,091	0,025	0,025	0,037	0,045	0,074
Total REE	0,18	0,53	6,66	4,57	1,25	1,51	2,31	7,43
$(La/Yb)_n$	3,69	3,52	0,44	1,82	1,1	N.d.	1,12	0,940
$(Eu/Eu^*)_n$	1,07	0,88	0,97	1,30	N.d.	»	N.d.	0,999
Rb	0,21*	0,36	1,11	0,77	»	»	»	0,635
Sr	0,33*	0,97	38	35	»	»	»	21,1
Y	0,051*	0,16	5,9	1,44	»	»	»	4,55
Zr	35	47	78	97	»	»	»	11,2
Nb	0,039*	0,14	0,10	0,032*	»	»	»	0,713
Ba	0,23	0,97	16,7	22	»	»	»	6,989
Hf	0,79	0,92	3,4	2,1	»	»	»	0,309
Th	0,028*	0,051	0,076	0,035*	»	»	»	0,085
U	0,010*	0,011*	0,036*	0,030*	»	»	»	0,021
Os	0,50	0,70	0,06	0,16	»	»	»	3,3
Ir	0,10	0,49	0,023	0,04	»	»	»	3,6
Ru	0,64	0,57	0,09	0,15	»	»	»	4,3
Pt	4,96	0,14	2,89	10,52	»	»	»	7,0
Pd	0,21	0,46	18,5	26,4	»	»	»	N.d.
Total PGE	6,41	2,36	21,6	37,2	»	»	»	22,2
Re	0,004*	0,006*	0,006*	0,015	»	»	»	N.d.
Pd/Ir	2,1	0,94	804	660	»	»	»	»
Pt/Ir	49,6	0,29	126	263	»	»	»	1,94

Note: Analyses were done in Analytical Center of IGM SB RAS, ICP-MS method (analysts S.V. Palessky, I.V. Nikolaeva, O.A. Koz'menko). *Elements composition is lower then the detection limits. ** REE composition in the primitive mantle - data [Sun, McDonough, 1989]; PGE composition – data [Rollinson, 1993].

Table 7. Average platinum group element and gold composition (ppb) of rocks from Naransky massif (Western Mongolia)

Rock	Pd	Pt	Rh	Au	Number of analyses
Dunites serpentinized	7,3	8,9	N.d.	14,7	6
Serpentinites apodunite	8,6	4,5	»	11,5	4
«	8,1	7,5	»	18,1	10
Wherlites	10,0 (4)	15,7 (4)	1,5 (1)	–	–
Websterites	12,4 (21)	12,9 (21)	1,7 (5)	22,0 (6)	–
Websterites Pl-bearing	21,8 (7)	14,0 (7)	N.d.	59,7 (5)	–
Gabbro-norites	24,0	8,7	»	77,6	5
Chromitites	10,7 (38)	22,0 (32)	8,2 (37)	Os – 58,3; Ir –123; Ru – 84,3	

Note. Analyses were done in Analytical Center of IGM SB RAS by atom-absorption method (analyst V.G. Tsimbalist).

In brackets – number of analysis.

It should be emphasized that, despite the overall depletion by REE the analyzed ultramafites of this massif are anomalously enriched with LREE (Figure 11). This kind of anomalous enrichment of ultramafic restites with LREE, according to our observations, is due to the presence of not only structural (isomorphic) impurity of these elements, but also of variable amounts of unstructured impurity in intragranular and intergranular microcracks. Total REE content in the gabbronorites of the massif (4.57 ppm) yields to its content in rocks from many other mafic-ultramafic massifs [Lesnov, 2010]. Possibly, the gabbronorites from the considered massif crystallized from the melt, which emerged during the partial melting of pre-depleted mantle source.

According to the determinations made by ICP-MS method, the total contents of Os, Ir, Ru, Pt and Pd in dunite and harzburgite are inferior to their total content in websterite and gabbro-norite. The content of Ir in the harzburgite was higher than in dunite. The content of Re in all these rocks is very low (Table 6). The results of PGE determination in rocks of the massif made by atomic absorption method are in good accordance with their estimations by ICP-MS method on Pt - in dunite and gabbro-norite, as well as on Pd - in websterite and gabbro-norite (Table 7). The chondrite-normalized patterns of distribution of the primitive mantle of contents of PGE and Re in websterite and gabbro-norite that were obtained by ICP-MS method are very similar by their configuration. Judging by them, in both of these rocks there is an insignificant Ir deficit, as well as some Pd excess (Figure 12). Studied rocks of this massif, particularly harzburgite, are characterized by low values of Ru/Ir parameter. The values of Pd/Ir parameter in websterite and gabbro-norite are two orders of magnitude higher than in the dunite and harzburgite.

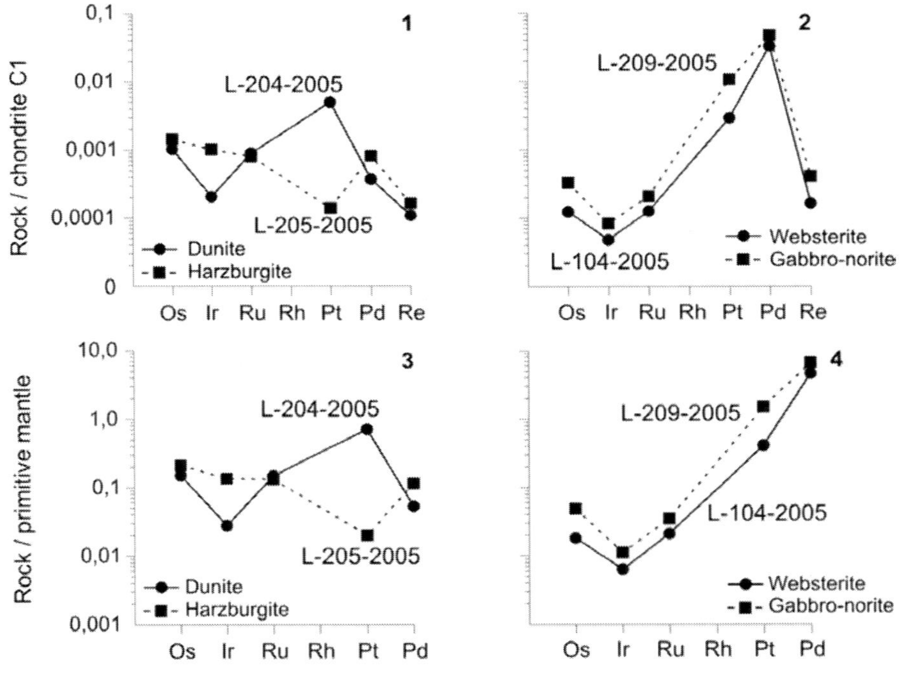

Figure 12. Chondrite- normalized (1, 2) and primitive mantle-normalized (3, 4) PGE patterns for rocks from Naransky massif (West Mongolia) (data Table 6).

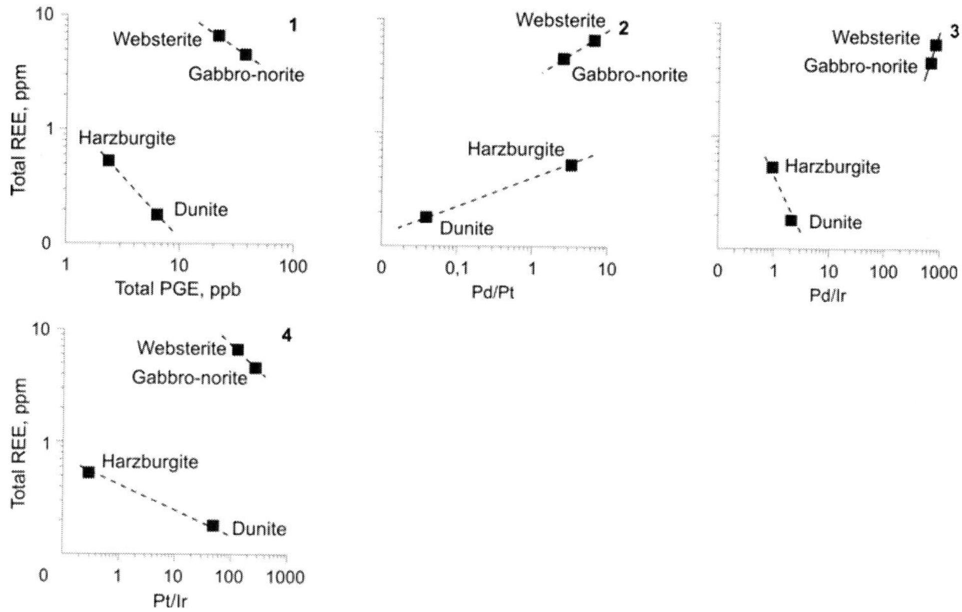

Figure 13. Graphs of correlation between total REE composition and total PGE composition (1), as well as Pd/Pt (2), Pd/Ir (3) and Pt/Ir (4) parameters values for rocks from Naransky massif (West Mongolia) (data Table 6).

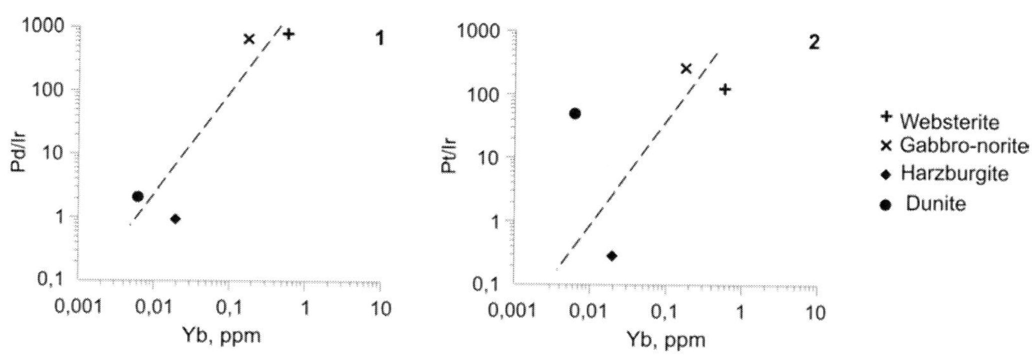

Figure 14. Graphs of correlation between Yb composition and Pd /Ir (1) and Pt/Ir (2) parameters values for rocks from Naransky massif (West Mongolia) (data Table 6).

The diagrams in the coordinates of total REE - total PGE and total REE - Pd/Pt the figurative points of dunite and harzburgite are located apart from the points of websterites and gabbronorites. (Figure 13, 1). On these diagrams, just like the diagram in the coordinates (total REE - Pd / Pt) we can assume the existence of two separate trends (Figure 13, 2-4). The position of figurative points in the diagrams in the coordinates (Yb - Pd / Ir) and (Yb - Pt / Ir) assumes discontinuity of geochemical characteristics of the ultramafites and gabbro. In these rocks with increasing contents of less refractory Pd and Pt against more refractory Ir contents, the contents of Yb are increasing (Figure 14). In general, these data do not contradict the assumption that during the formation of ultramafic restites and gabbros composing this massif, a multidirectional fractionation of REE and PGE occurred.

Kokpektinsky massif (South Urals, Kazakhstan) is located in the basin of the Kokpekty River (South Urals) and structurally confined to the area of Main Urals deep-seated fault [Petrology..., 1991; Fershtater & Bea, 1996; Fershtater, 2004]. It is exposed over an area of 70 sq. km. and has a form that is elongated to north-eastern direction.

This massif is considerably gabbro and is directly connected with the large Kempirsaysky ultramafic protrusion located to the east of it; they both comprise the Kempirsay ophiolite association. The structure of Kokpektinsky massif includes troctolites, olivine gabbro and gabbro, and also the located among them isolations of serpentinized dunites and wehrlites of various sizes.

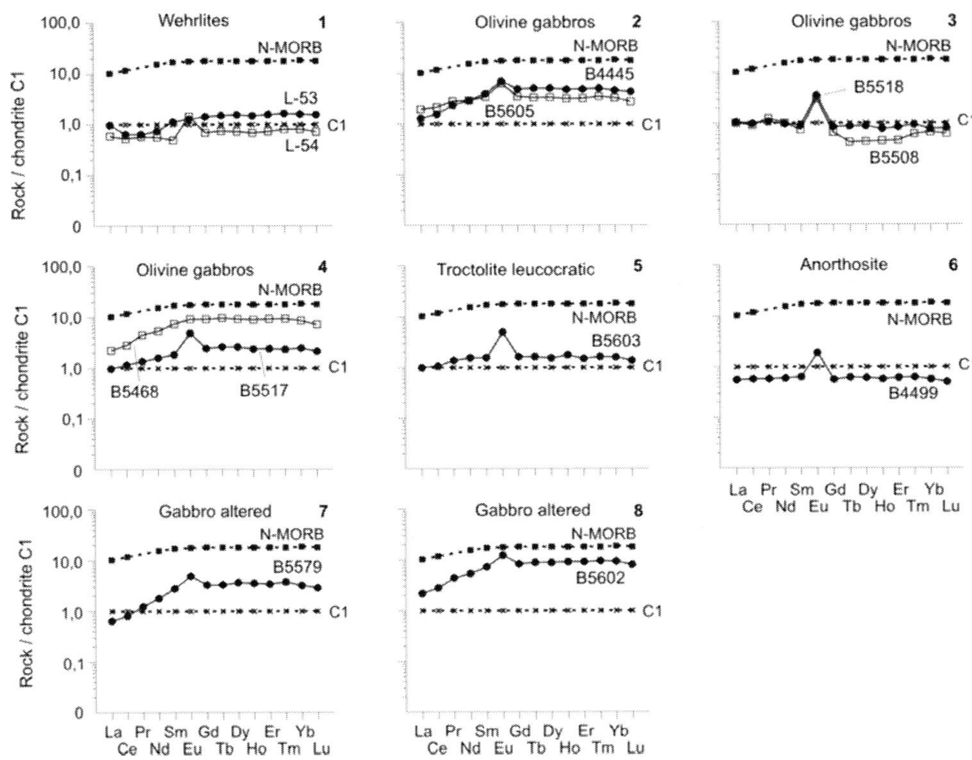

Figure15. Chondrite-normalized REE patterns for rocks from Kokpektinsky massif (South Urals, Kazakhstan) (data Table 8). N-MORB – (by [Sun, McDonough, 1989].

The rocks of the massif are cut by dikes of pegmatite gabbro and anorthosites. Among the gabbros of Kokpektinsky massif close to the contact with Kempirsaysky ultramafic protrusion there were xenoliths of ultramafites forming this protrusion found. The xenoliths size varies from a few centimeters to hundreds of meters, the shape - from angular to round and the contacts with others gabbros - from sharp to gradual.

According to our observations, the Kokpektinsky massif has intruded after the formation of Kempirsaysky ultramafic protrusion [Petrology ..., 1991], which gives grounds to consider it as a polygenic ultramafic-mafic body. It is assumed that whole variety of rock types that compose the massif was formed as a result of crystallization of primary mantle, as well as contaminated mafic melts that appeared under the active collaboration of primary melts with ultramafic restite substance.

According to the results of dating by different isotopic methods (Rb-Sr, Sm-Nd, U-Pb, Ar-Ar) it was ascertained that Kempirsay ophiolite association was formed in the range from 420 to 396 Ma [Edwards & Wasserburg, 1985; Sharma & Wasserburg, 1996; Sharma et al., 1995; Melcher et al., 1999].

Judging by the data obtained by ICP-MS method, the rocks of Kokpektinsky massif are largely depleted by REE, the chondrite-normalized contents of which are generally lower than in N-MORB basalts (Table 8, Figure 15). As we can see by the level of REE accumulation, the wehrlites and anorthosite are inferior to non-olivine and olivine gabbro, while the chondrite-normalized Yb contents in them are slightly higher than La contents.

Note that in a sample of anorthosite there is no significant excess of chondrite-normalized LREE contents over HREE observed, as typical of many samples enriched with plagioclase.

In addition, in the rocks of this massif, especially in the olivine gabbro and troctolites, there is a slight Eu excess, and the lowest values of $(Eu/Eu^{*})_n$ parameter were identified in wehrlites, while the highest - in the olivine gabbro. By the level of REE accumulation, as well as by the configuration of their distribution patterns the rocks of Kokpektinsky massif are in many respects comparable to the same samples from Samail massif composing the ophiolite association of Oman [Pallister, Knight, 1981].

In the rocks of Kokpektinsky massif analyzed by ICP-MS, the total Os, Ir, Ru, Pt and Pd contents range from 0.6 to 18,5 ppb, the highest total PGE content was determined for one of the gabbro samples that was subjected to epigenetic changes (Table 8). Values of Pd/Pt parameter in rocks of the massif range in between 0.39-3.09. Almost all of the rocks from the massif are depleted by Re. All the rocks of the massif, excepting anorthosite sample, have the patterns of the distribution of PGE and Re with a steep positive slope and some of them are complicated by a minimum of Ru (Figure 16). Figurative points of the composition of the rocks from the massif in the diagrams in the coordinates of (Pt-Ir) and (Pd -Ir) form two distinct aggregates, corresponding to two types of them - more depleted and less depleted with respect to Pt and Pd. The second type includes mainly olivine gabbro. In the diagram in the coordinates of Pd/Ir-$(La/Yb)_n<1$ the points corresponding to the composition of gabbro, part of olivine gabbro samples, of troctolites, anorthosites and wehrlites, are located along two weakly manifested trends with negative slopes; another trend is formed by the points of olivine gabbro samples that have the values of parameter $(La/Yb)_n>1$. In addition, figurative points of the massif rocks form two poorly defined trends with a negative slope in the diagram in the coordinates (Pt - HREE), while the points corresponding to the composition of wehrlites and anorthosites are away from these trends (Figure 17). The values of Pd/Ir and Pt/Ir parameters, as well as the content of Yb in these rocks are subject to considerable variation, but no relationship between these characteristics was detected (Table 8, Figure 18).

Thus, the data on the geochemistry of REE and PGE suggests that in the rocks from Kokpektinsky massif there is an inverse relationship between the contents of elements of these two groups. It is not excluded, however, that in those rocks that presumably occurred during the crystallization of mantle mafic melts contaminated by a substance of older ultramafic restites, i.e. have a hybrid nature (olivine gabbro, troctolites, wehrlites), the originally existed inverse relationship between the contents of REE and PGE could be disturbed.

Table 8. Rare earth element, rare element (ppm), platinum group element, and rhenium (ppb) composition of rocks from Kokpektinsky massif (South Urals, Kazakhstan) [Lesnov, Balykin, 2007]

Element	L-53	L-54	B4445	B5605	B5518	B5508	B5517	B5468	B5603	B4499	B5579	B5602
	Olivine gabbros								*Troctolite leucocratic*	*Anorthosite*	*Gabbros altered*	
La	0,24	0,14	0,32	0,48	0,26	0,25	0,24	0,55	0,24	0,14	0,16	0,53
Ce	0,41	0,33	0,99	1,37	0,63	0,59	0,74	1,81	0,68	0,37	0,52	1,78
Pr	0,057	0,051	0,21	0,25	0,096	0,11	0,12	0,40	0,12	0,052	0,11	0,39
Nd	0,35	0,27	1,39	1,39	0,46	0,48	0,76	2,6	0,74	0,28	0,85	2,5
Sm	0,17	0,076	0,61	0,53	0,14	0,11	0,28	1,15	0,24	0,098	0,43	1,12
Eu	0,075	0,085	0,41	0,37	0,21	0,18	0,28	0,54	0,29	0,11	0,28	0,71
Gd	0,29	0,14	1,01	0,72	0,17	0,13	0,50	1,88	0,33	0,12	0,67	1,70
Tb	0,057	0,028	0,19	0,13	0,033	0,016	0,098	0,36	0,061	0,023	0,12	0,33
Dy	0,40	0,19	1,29	0,86	0,22	0,11	0,67	2,3	0,39	0,16	0,93	2,2
Ho	0,084	0,039	0,27	0,18	0,044	0,025	0,14	0,51	0,10	0,033	0,20	0,52
Er	0,26	0,12	0,81	0,53	0,14	0,076	0,40	1,54	0,25	0,10	0,57	1,51
Tm	0,043	0,021	0,13	0,090	0,024	0,016	0,060	0,24	0,042	0,016	0,096	0,24
Yb	0,27	0,13	0,76	0,55	0,13	0,11	0,41	1,42	0,26	0,094	0,53	1,53
Lu	0,040	0,018	0,11	0,070	0,020	0,016	0,055	0,18	0,035	0,013	0,073	0,20
Total REE	2,75	1,65	8,51	7,52	2,57	2,22	4,76	15,5	3,79	1,60	5,53	15,4
(La/Yb)*n*	0,60	0,73	0,28	0,59	1,38	1,52	0,39	0,26	0,62	0,98	0,20	0,23
(Eu/Eu*)*n*	1,01	2,45	1,57	1,80	4,02	4,42	2,28	11,1	3,13	3,18	1,62	1,57
Rb	0,33	0,58	0,59	0,75	0,78	0,83	0,93	0,74	0,26	0,55	1,73	1,38
Ba	2,4	20 3,	3,3	5,4	5,1	15,3	2,2	4,3	5,3	3,2	14,0	9,2
Sr	24	59	145	176	182	109	127	145	162	115	113	122
Th	0,010	0,025	0,009	0,046	0,014	0,013	0,007	0,021	0,010	0,017	0,011	0,015
U	0,032	0,028	0,027	0,026	0,037	0,084	0,016	0,026	0,013	0,032	0,024	0,058
Nb	0,080	0,095	0,083	0,23	0,087	0,069	0,11	0,21	0,051	0,099	0,22	0,10
Zr	55	68	171	194	179	96	293	409	159	173	172	158
Hf	2,5	3,1	7,5	8,5	7,7	3,3	12,9	17,9	7,4	7,4	7,8	7,3
Y	2,6	1,28	8,6	6,2	1,94	0,72	4,7	16,8	2,6	1,16	6,1	14,8
Os	N.d.	N.d.	N.d.	0,13	0,1	N.d.	0,07	0,13	0,06	0,15	0,12	0,05
Ir	0,05	0,06	0,03	0,11	0,06	0,06	0,11	0,05	0,05	0,04	0,11	0,02
Ru	0,16	0,28	0,01	0,23	0,01	N.d.	N.d.	0,01	0,02	N.d.	0,06	0,11
Pt	0,36	0,23	0,33	2,89	2,68	1,96	1,02	1,68	1,97	0,12	9,06	0,3
Pd	0,85	0,71	0,36	3,19	2,31	2,58	0,47	2,84	0,76	0,29	9,18	0,44
Total PGE	1,42	1,28	0,73	6,55	5,16	5,60	1,67	4,71	2,86	0,60	18,53	0,92
Re	0,08	0,35	0,24	0,89	0,35	0,16	0,26	0,05	0,38	0,02	0,13	0,21
Pd/Pt	2,4	3,1	1,1	1,1	0,86	1,3	0,46	1,7	0,39	2,4	1,0	1,5
Pd/Ir	17,0	11,8	12,0	29,0	38,5	43,0	4,3	56,8	15,2	7,3	83,5	22,0
Pt/Ir	7,2	3,8	11,0	26,3	44,7	32,7	9,3	33,6	39,4	3,0	82,4	15,0
Ru/Ir	3,2	4,7	0,33	2,1	0,17	N.d.	N.d.	0,20	0,40	N.d.	0,55	5,5

Note. Analysis were done in Analytical Center of IGM SB RAS, ICP-MS method (analysts S.V. Palessky, I.V. Nikolaeva, O.A. Koz'menko).

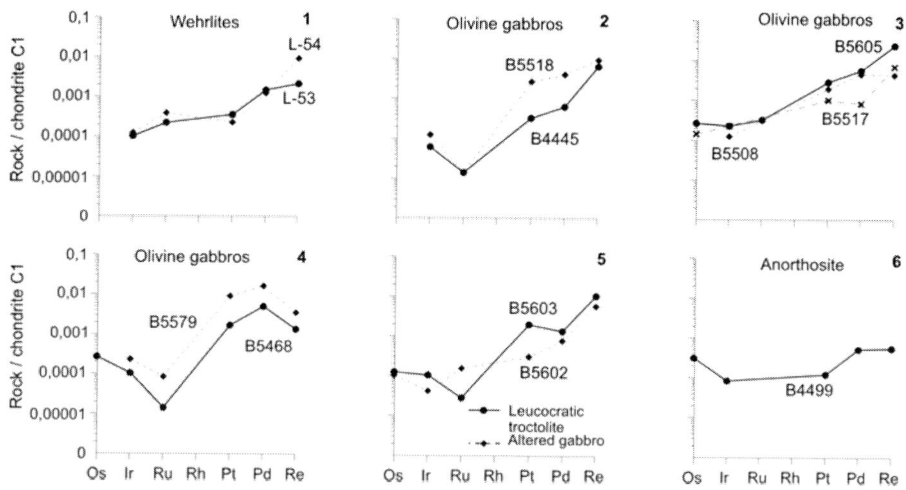

Figure16. Chondrite-normalized PGE and Re patterns for rocks from Kokpektinsky massif (South Urals, Kazakhstan) (data Table 8).

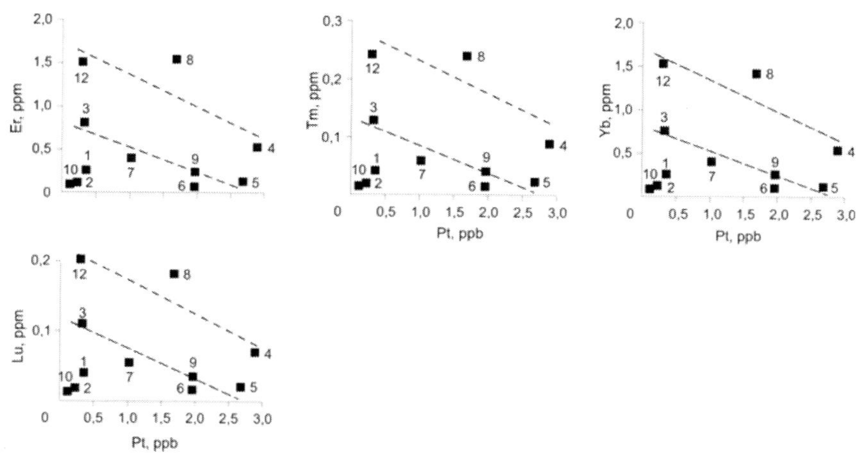

Figure 17. Graphs of correlation between Pt composition and heavy REE composition for rocks from Kokpektinsky massif (South Urals, Kazakhstan). Numbers of points correspond to the numbers of samples in the Table 8.

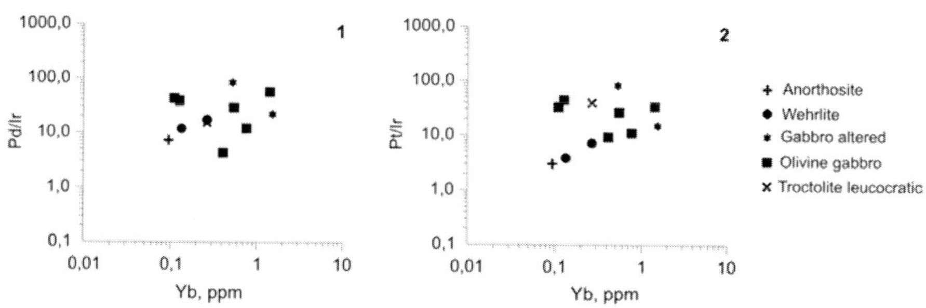

Figure 18. Graphs of correlation between Yb composition and Pd/Ir (1) and Pt/Ir (2) parameters values for rocks from Kokpektinsky massif (South Urals, Kazakhstan) (data Table 8).

Ultramafic Xenoliths in Alkaline Basalts from Eifel and Vogelsberg Provinces (Germany)

Table 9. Rare earth element (ppm), platinum group element, and gold (ppb) composition of harzburgites, lherzolites, and pyroxenites from xenoliths in Eifel and Vogelsberg alkaline-basalt provinces (Germany)

Element	194*	826	832	589*	1343*	1342	211*	229*	1317*
			Harzburgites						*Lherzolite*
La	0,92	0,30	0,42	1,68	1,86	0,38	3,30	2,07	2,51
Ce	2,40	0,41	0,58	5,00	7,00	0,87	10,20	5,30	9,20
Nd	0,90	N.d.	0,37	N.d.	3,30	N.d.	3,80	3,00	2,70
Sm	0,25	0,07	0,10	0,51	0,62	0,10	0,82	0,43	0,63
Eu	0,07	0,02	0,03	0,15	0,17	0,03	0,22	0,13	0,18
Tb	0,03	0,03	0,03	0,05	0,06	N.d.	0,07	0,06	0,08
Yb	0,07	0,04	0,06	0,11	0,12	0,05	0,15	0,12	0,20
Lu 0,01		0,01	0,02	0,02	0,02	0,01	0,02	0,02	0,04
Total REE	4,65	N.d.	1,60	N.d.	13,15	N.d.	18,59	11,12	15,54
(La/Yb)*n*	9,27	5,06	4,80	10,3	10,5	5,61	14,7	11,6	8,51
Os	0,28	1,30	0,18	4,24	4,55	1,34	6,47	1,37	3,37
Ir	0,36	2,19	0,58	7,06	5,46	1,06	9,07	1,91	3,36
Ru	1,70	2,82	2,74	14,48	8,22	1,62	15,98	2,93	7,16
Rh	0,47	0,43	0,41	1,48	2,94	0,73	3,23	1,01	1,69
Pt	1,83	2,80	1,80	13,51	12,40	1,65	18,00	4,55	12,17
Pd	1,22	1,11	1,41	7,30	5,74	0,41	8,42	1,98	0,42
Au	0,18	0,43	0,50	0,66	0,79	0,25	0,10	0,35	0,12
Total PGE+Au	6,04	11,08	7,62	48,73	40,10	7,06	61,27	14,10	28,29
Pd/Ir	3,39	0,51	2,43	1,03	1,05	0,39	0,93	1,04	0,13
Pt/Ir	5,08	1,28	3,10	1,91	2,27	1,56	1,98	2,38	3,62

The contents of REE and PGE were determined in non-amphibole and amphibole harzburgites, lherzolites and pyroxenites, which composed xenoliths in alkaline basalts from these provinces [Schmidt et al., 2003]. It was ascertained that in most cases the contents of LREE, as well as total REE contents, in non-amphibole samples of peridotites are lower than their contents in the amphibole samples (Table 9). Total PGE and Au contents in non-amphibole peridotites were also slightly lower than in their amphibole samples. In addition, it was determined that the REE content in pyroxenites is higher, while PGE content is lower than in harzburgites and lherzolites. In the diagrams in the coordinates (La-PGE) and (Sm-PGE) the figurative points corresponding to the composition of harzburgites and lherzolites form positively sloped trends, while in the diagrams in the coordinates (Yb-PGE) correlation between these characteristics is not observed (Figure 19).

These data suggest that in the rocks from these xenoliths the La and other LREE, as well as some PGE, had undergone a redistribution of different intensity and trend that occurred during the infiltration of fluid components coming from alkaline basalts relocating them. Apparently, the less mobile HREE in contrast to the highly mobile LREE were subject to such redistribution to a lesser extent. In addition, it is likely that those amphiboles that occur in these ultramafites could be formed under the influence of the same fluids. Judging by the location of the figurative points in the diagrams in the coordinates (Yb-Pd/Ir) and (Yb-Pt/Ir)

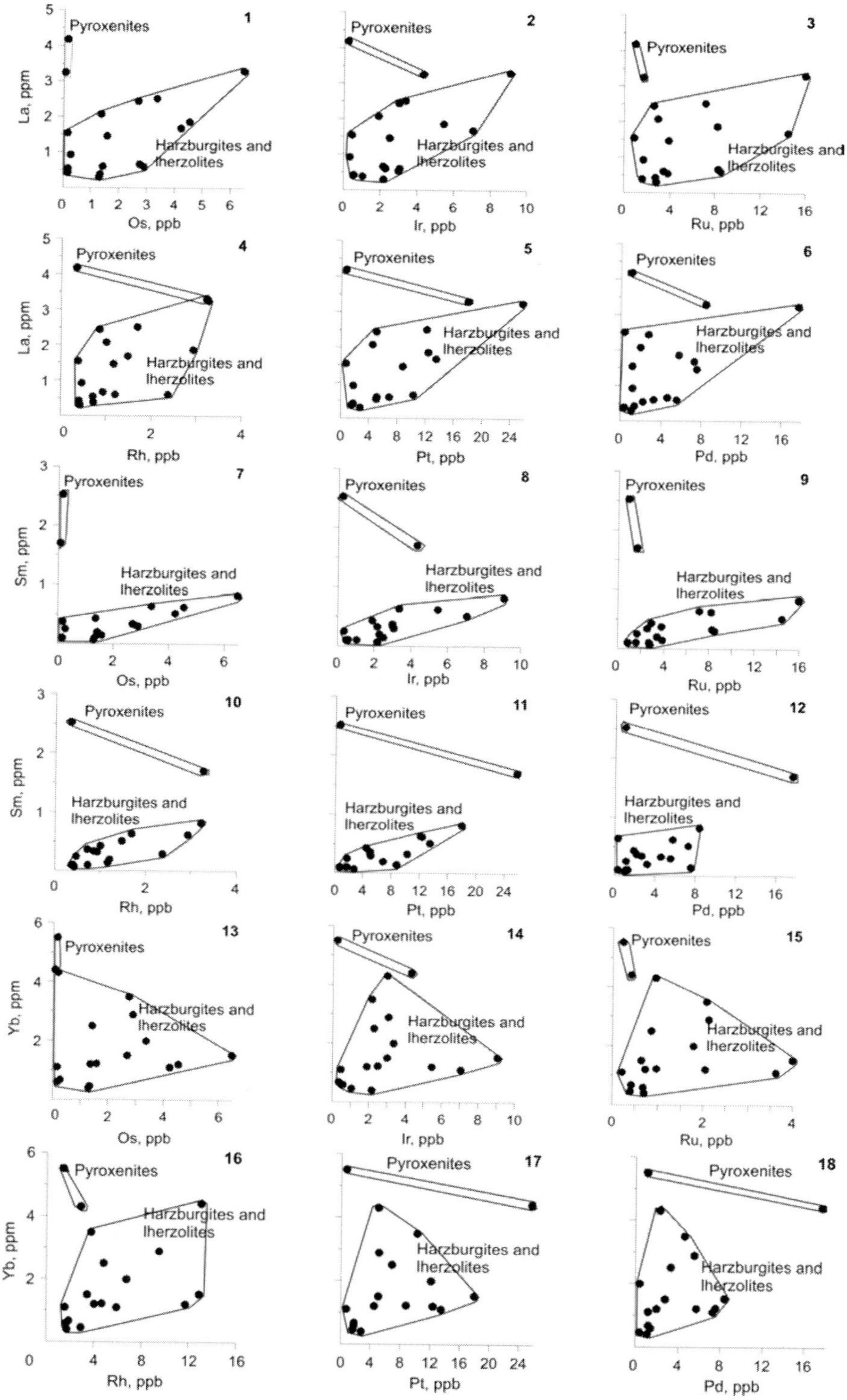

Figure 19. Graphs of correlation between PGE composition and La (1-6), Sm (7-12), Yb (13-18) composition for rocks from xenoliths in alkali basalts of Eiffel and Vogelsberg provinces (Germany) (data Table 9).

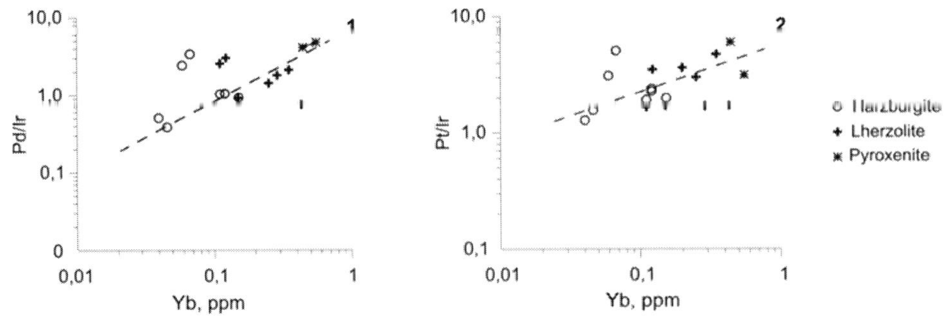

Figure 20. Graphs of correlation between Yb composition and Pd/Ir (1) and Pt/Ir (2) parameters values for rocks from xenoliths in alkali basalts of Eifel and Vogelsberg provinces (Germany) (data Table. 9).

Table 10. Rare earth element (ppm), and platinum group element (ppb) composition of dunites, and harzburgites from Luobusa massif (South Tibet, China)

	[Zhou et al., 2005], ICP-MS											
Element	D0	D5	D10	D15	D20	D25	D30	D55	D65	D70	D75	D80
	Dunites											
La	9,88	5,70	7,94	6,49	9,91	9,76	2,13	1,89	2,61	2,21	3,62	7,82
Ce	18,0	10,4	14,5	17,8	15,8	15,6	3,17	2,90	3,84	35,7	5,44	13,3
Pr	2,13	1,18	1,28	1,33	2,10	17,6	0,41	0,33	0,46	0,44	0,58	1,35
Nd	8,10	4,79	3,38	2,46	5,87	4,64	1,28	1,16	1,41	1,26	1,76	4,12
Sm	1,60	0,85	0,94	0,27	1,28	1,23	0,26	0,23	0,20	0,20	0,34	0,69
Eu	0,41	0,28	0,20	0,20	0,21	0,18	0,15	0,15	0,17	0,13	0,24	0,35
Gd	1,39	1,10	1,54	0,61	1,93	1,44	0,38	0,42	0,51	0,28	0,42	1,30
Tb	0,27	0,23	0,44	0,11	0,38	0,29	0,08	0,01	0,13	0,06	0,14	0,31
Dy	2,07	2,09	3,46	1,00	2,97	2,49	0,88	0,91	1,48	0,42	1,55	3,08
Ho	0,68	0,72	1,25	0,33	0,95	0,75	0,32	0,44	0,63	0,22	0,57	1,00
Er	2,86	2,99	4,76	1,76	3,58	3,32	1,64	1,87	3,25	1,09	3,13	4,24
Tm	0,72	0,70	1,00	0,43	0,78	0,78	0,41	0,53	0,80	0,38	0,70	0,98
Yb	6,58	6,55	9,21	4,49	7,30	6,66	4,40	5,41	7,60	4,83	7,33	8,62
Lu	1,41	1,43	1,84	1,09	1,47	1,49	1,03	1,14	1,73	1,34	1,67	1,91
Total REE	56,09	39,02	51,74	38,41	54,55	66,22	16,54	17,39	24,82	48,56	27,49	49,07
(La/Yb)*n*	1,01	0,59	0,58	0,98	0,92	0,99	0,33	0,24	0,23	0,31	0,33	0,61
Ir	5,72	6,16	5,13	6,32	5,11	5,33	4,54	5,24	4,17	5,01	5,77	4,63
Ru	10,8	7,72	8,41	10,7	7,12	10,3	10,0	9,8	7,41	8,26	8,92	8,62
Rh	0,75	0,64	1,57	1,06	0,52	0,42	0,73	1,09	0,52	0,79	0,59	1,42
Pt	2,02	1,61	9,67	3,17	1,18	1,61	1,84	3,99	2,10	1,91	1,95	9,02
Pd	3,23	3,57	4,93	3,04	1,68	1,99	2,97	3,06	2,40	1,92	1,73	5,13
Total PGE	22,52	19,7	29,71	24,29	15,61	19,67	20,12	23,18	16,6	17,89	18,96	28,82
Rh+Pt+Pd	6,00	5,82	16,2	7,27	3,38	4,02	5,54	8,14	5,02	4,62	4,27	15,57
Pd/Ir	0,56	0,58	0,96	0,48	0,33	0,37	0,65	0,58	0,58	0,38	0,30	1,1
Pt/Ir	0,35	0,26	1,88	0,50	0,23	0,30	0,41	0,76	0,50	0,38	0,34	2,0

in the row from harzburgites to lherzolites and pyroxenites, the values of these three characteristics are increasing (Figure 20). Thus, we believe that under the influence of fluids coming from basaltic melts into the xenoliths of ultramafites, not only LREE, but also Pd and Pt could change to a relatively mobile state. If this assumption is confirmed on the basis of representative analytical data, it will serve as a convincing argument in favor of the view that the results of REE and PGE analyses in ultramafites from upper mantle xenoliths in most cases do not correspond to those contents of these elements, which originally characterized them; that is, till the moment when these ultramafites came into direct contact with the basaltic melts carrying them. At the same time, these observations give grounds to believe that the infiltration of basalt fluids has been able to more or less disturb the original inverse relationship between the contents of REE and PGE in the ultramafic restites.

Luobusa massif (Southern Tibet, China) that is a part of ophiolite association of Southern Tibet is a steeply dipping lenticular body having a length of 20 km. and a width of 4 km. The massif is composed of harzburgites and subordinate dunites [Zhou et al., 2005]. Dunites form a large separate body lying along a tectonic contact of the massif with host rocks, as well as several small lenticular isolations occurring among harzburgites. In addition, the dunites are composing narrow zones around the bodies of ore chromitites localized among harzburgites.

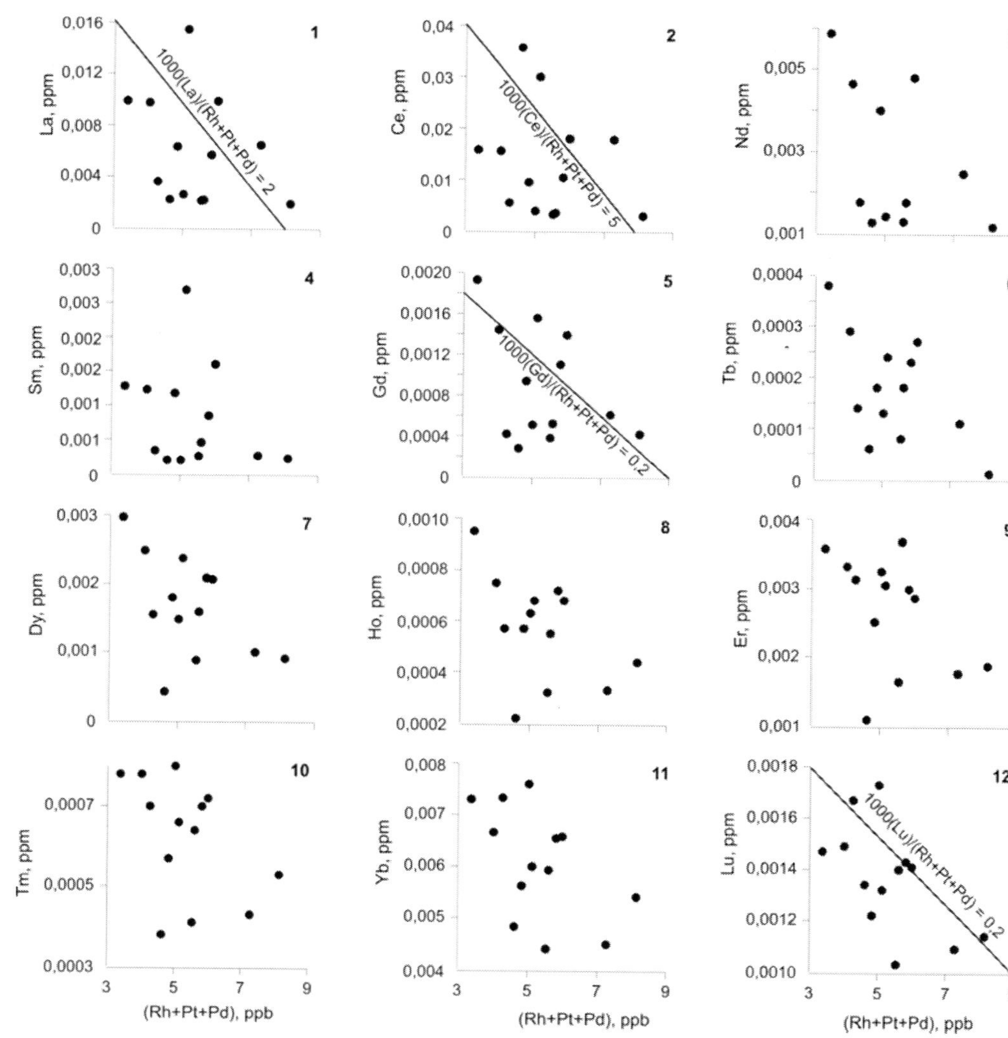

Figure 21. Graphs of correlation between REE composition and Rh, Pt and Pd total composition for dunites from Luobusa massif (Tibet, China) (data Table 10).

As follows from the analysis, the content of REE in harzburgites is usually slightly higher than in dunites, with the exception of those types of dunites that form isolations among harzburgites (Table 10). Total REE contents in the dunites from their large body range from 17 to 75 ppm and in those samples that compose small isolations among harzburgites, it is higher and the average is 114 ppm. In harzburgites the total REE content ranges from 280 to 1250 ppm. Both among the dunites and harzburgites there are some samples with abnormally high LREE content.

Total Ir, Ru, Rh, Pt and Pd contents in ultramafites of this massif range from 16 to 29 ppb, while the configuration of patterns of PGE distribution show some Pt deficiency in dunites. The presented graphs indicate that dunites have an inverse relationship between total Pd, Pt and Rh contents on the one hand, and the contents of REE on the other hand (Figure 21). One can also assume that in the dunites and harzburgites from the massif there is an inverse relationship between Yb contents and the values of Pd/Ir and Pt/Ir parameters (Figure 22). In general, data on the distribution of REE and PGE in dunites and harzburgites from Luobusa ultramafic massif do not contradict the model of their formation that occurred during the partial melting of mantle source.

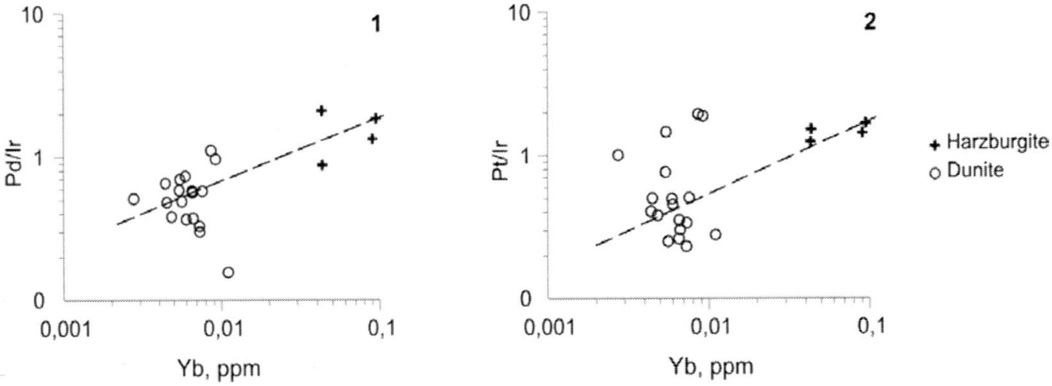

Figure 10.22. Graphs of correlation between Yb composition and Pd/Ir (1) and Pt/Ir (2) parameters values for dunites from Luobusa massif (Tibet, China) (data Table 10.10).

Iti and Kalidromon Massifs (Greece)

These two massifs that are composed of lherzolites and subordinate harzburgites are a part of the ophiolite association of Central Greece [Karipi et al., 2006]. In lherzolites from both massifs the total REE content ranges from 0.83 to 1.81 ppm, in harzburgites - from 0.47 to 0.57 ppm (Table 11).

Total PGE content in lherzolites ranges from 15.1 to 38.8 ppb, which is slightly higher than in harzburgites (9.6-18.1 ppb). By the position of figurative points on the graphs we can assume that in these ultramafites the contents of Sm, Eu and Gd are inversely related to the contents of Os and that for some reason there is a direct relation between REE and Ir (Figure 23). There is also an inverse relation between Yb content and the values of Pd/Ir and Pt/Ir parameters (Figure 24).

Table 11. Rare earth element (ppm), platinum group element, and gold (ppb) composition of rocks from Iti and Kallindromon massifs (Greece)

[Karipi et al., 2006], INN										
Iti					Kallindromon				Iti	
Element	I-126	I-130	I-617	I-290	K-119	K-124a	K-144	K-124	I-189	I-317
Lherzolites									*Harzburgites*	
La	N.d.	0,06	N.d.	N.d.	N.d.	N.d.	N.d.	N.d.	0,25	0,28
Ce	»	0,08	»	0,05	»	0,07	0,06	0,12	0,22	0,29
Pr	0,02	N.d.	0,01	N.d.	»	0,02	0,02	0,01	0,04	0,04
Nd	0,18	0,07	0,09	0,06	0,16	0,12	0,14	N.d.	0,1	0,12
Sm	0,13	0,09	0,05	0,05	0,07	0,05	0,06	0,05	0,03	0,03
Eu	0,06	0,05	0,03	0,03	0,03	0,02	0,03	0,02	0,01	0,01
Gd	0,24	0,20	0,11	0,09	0,12	0,1	0,12	0,1	0,02	0,02
Tb	0,05	0,05	0,03	N.d.	0,03	0,02	0,03	N.d.	N.d.	N.d.
Dy	0,37	0,31	0,22	0,20	0,21	0,19	0,21	0,19	0,02	0,03
Ho	0,09	0,07	0,05	0,05	0,05	0,05	0,05	0,05	N.d.	N.d.
Er	0,27	0,20	0,17	0,18	0,16	0,16	0,17	0,16	0,03	0,04
Tm	0,05	0,03	0,03	0,03	0,03	0,03	0,03	0,03	0,01	0,01
Yb	0,3	0,18	0,18	0,17	0,18	0,19	0,19	0,19	0,16	0,20
Lu	0,05	0,03	0,03	0,04	0,03	0,03	0,03	0,03	0,05	0,07
Total REE	1,81	1,28	1,00	0,90	1,07	0,98	1,08	0,83	0,47	0,57
Os	2,0	2,0	2,0	2,0	2,3	2,0	1,9	2,5	3,0	2,0
Ir	3,3	2,2	2,3	1,5	3,3	2,8	2,7	3,2	1,9	0,2
Ru	6,5	6,2	5,3	N.d.	10,6	7,5	8,6	6,2	N.d.	N.d.
Rh	1,6	1,3	1,4	0,6	2,7	1,4	1,7	1,5	0,7	»
Pt	6,5	4,8	6,3	7,0	14,7	6,7	N.d.	7,6	N.d.	0,3
Pd	4,0	3,4	4,0	4,0	5,2	5,9	6,0	3,2	4,0	15,6
Total PGE	23,9	19,9	21,3	15,1	38,8	26,3	N.d.	24,2	N.d.	N.d.
Au	3,2	2,7	4,6	1,0	7,3	1,9	9,7	2,5	1,2	3,5
Pd/Ir	1,2	1,6	1,7	2,7	1,6	2,1	2,2	1,0	2,1	77
Pt/Ir	2,0	2,2	2,7	4,7	4,5	2,4	N.d.	2,4	N.d.	1,5

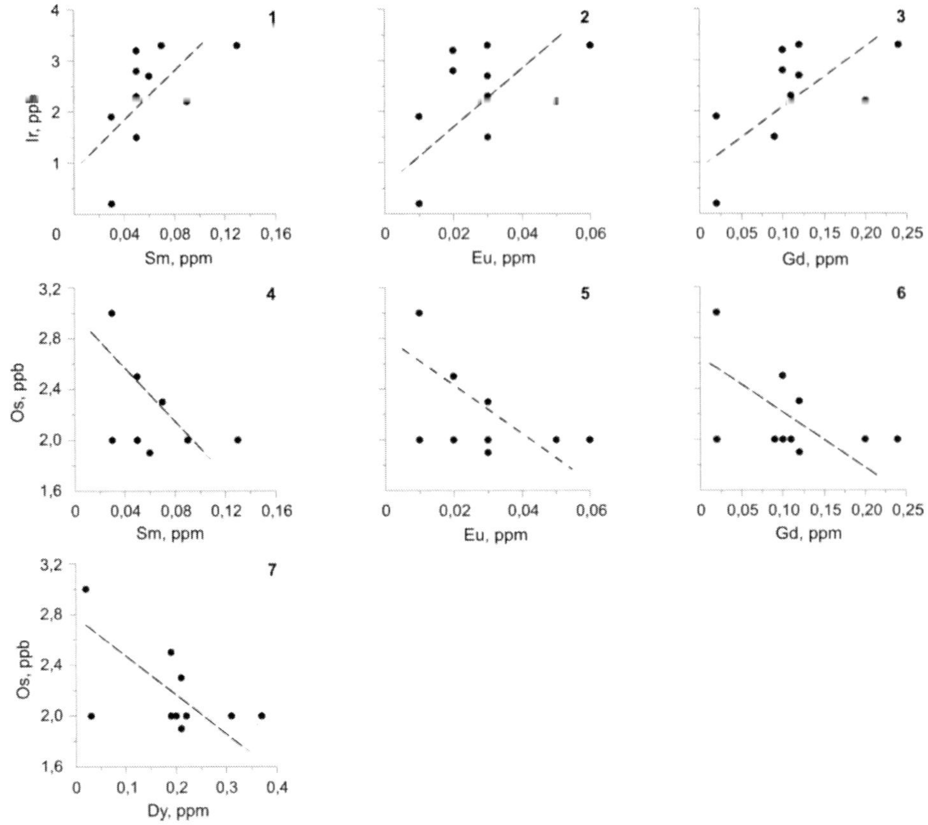

Figure 23. Graphs of correlation between Ir composition and Sm, Eu, Gd composition (1-3), as well as between Os composition and Sm, Eu, Gd, Dy composition (4-7) for lherzolites and harzburgites from Kallidromon and Iti massifs (Greece) (data Table 11).

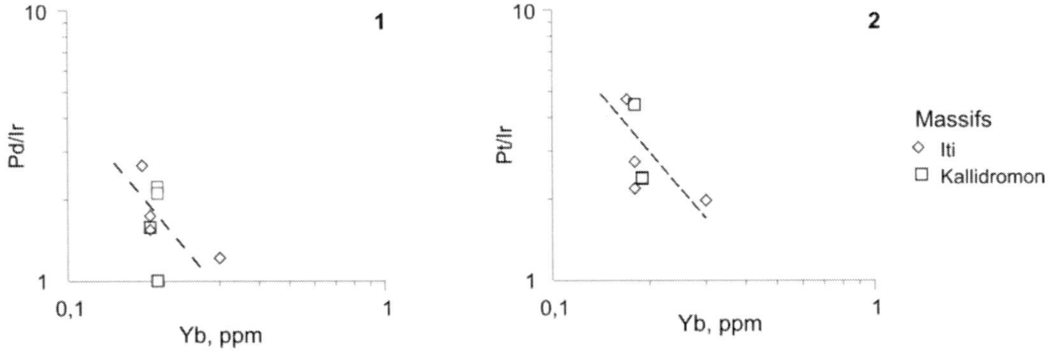

Figure 24. Graphs of correlation between Yb composition and Pd/Ir (1) and Pt/Ir (2) parameters values for lherzolites from Kallidromon and Iti massifs (Greece) (data Table 11).

KOMATIITE-BASALT COMPLEX OF NORTH-WEST VIETNAM

During the petrological and geochemical studies of the Permian-Triassic volcanic complex of North-West Vietnam in the composition of the residing komatiites, komatiitic and normal basalts the contents were determined for REE and among PGE – only for Os [Hanski et al., 2004; Lesnov, Balykin, 2005] (Table 12). Komatiites and komatiitic basalts have higher contents of MgO and Cr, higher contents of Os, as well as lower total REE contents compared to normal basalts. As can be seen in the diagrams, between the contents of MgO and Cr, as well as between MgO and Os, there is an evident direct relationship (Figure 25). Based on these data, we can assume that in the volcanic rocks of this complex, the parent melts of which were, obviously, formed at different degrees of partial melting of mantle sources, there is an inverse relation between REE contents and contents of all PGE.

Table 12. Rare earth element (ppm), Os (ppb), and MgO (wt %) composition of komatiites, basaltic komatiites, and tholeiite basalts (North-West Vietnam)

[Hanski et al., 2004; Balykin et al., 2004], ICP-MS								
Element	VB6892	VB6889	G1448	VB6891	V8/86	VB6865	V9/86	G1456
Komatiites								
La	0,68	0,64	0,91	1,24	0,66	0,36	1,00	0,70
Ce	2,03	2,05	2,90	3,20	2,05	1,18	2,39	2,08
Pr	0,38	0,39	N.d.	0,50	0,40	0,22	0,35	0,33
Nd	2,34	2,37	2,80	2,70	2,45	1,36	1,82	2,25
Sm	1,07	1,11	1,10	1,08	1,01	0,35	0,71	0,99
Eu	0,45	0,48	0,47	0,47	0,49	0,30	0,33	0,38
Gd	1,93	1,84	1,70	1,79	1,74	1,12	1,25	1,78
Tb	0,36	0,36	0,30	0,33	0,32	0,23	0,24	0,31
Dy	2,35	2,33	N.d.	2,23	2,17	1,58	1,67	2,39
Ho	0,51	0,53	»	0,51	0,45	0,37	0,37	0,49
Er	1,50	1,55	»	1,44	1,29	1,14	1,01	1,48
Tm	0,22	0,22	»	0,19	0,19	0,17	0,16	0,2
Yb	1,42	1,41	1,31	1,36	1,23	1,07	1,02	1,43
Lu	0,20	0,21	0,20	0,19	0,19	0,17	0,16	0,23
Total REE	15,4	15,5	N.d.	17,2	14,6	9,62	12,5	15,0
$(La/Yb)_n$	0,32	0,31	0,47	0,62	0,36	0,23	0,66	0,33
Cr	1844	1930	2300	1763	1804	2519	2575	2034
Os	3,915	2,841	2,626	2,529	2,511	2,082	1,874	1,625
MgO	22,37	22,08	20,53	21,05	22,13	22,6	24,85	20,78

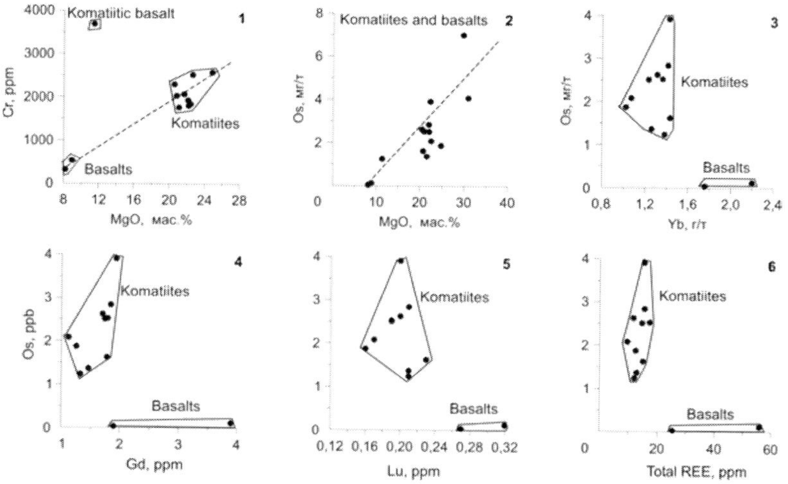

Figure 25. Graphs of correlation between Os composition and MgO (1), Cr (2), Yb (3), Gd (4), Lu (5) composition and total REE composition (6) for komatiites, komatiite basalts and tholeiite basalts from komatiite-basalt complex (North-West Vietnam) (data Table 12).

Komatiite-Basalt Complexes of Gorgona Island (Colombia), Alexo (Ontario, Canada), Onverwacht, Barberton, Perseverance (Africa) and Kambalda (West Australia) Provinces

Based on the analytical data obtained by geochemical studies of several komatiitic provinces, we can conclude that in these rocks between the contents of Os, Ir and Ru, on the one hand, and MgO contents, on the other hand, there is a direct relation, while between the contents of Rh, Pt and Pd and MgO contents - is an inverse relation [Brugmann et al., 1987]. At the same time, the position of figurative points on the presented diagrams shows that between MgO contents and contents of Dy and Yb there is an inverse relation (Figure 26). Thus, these facts also suggest, albeit indirectly, that in komatiites there is an inverse relation between REE and PGE contents.

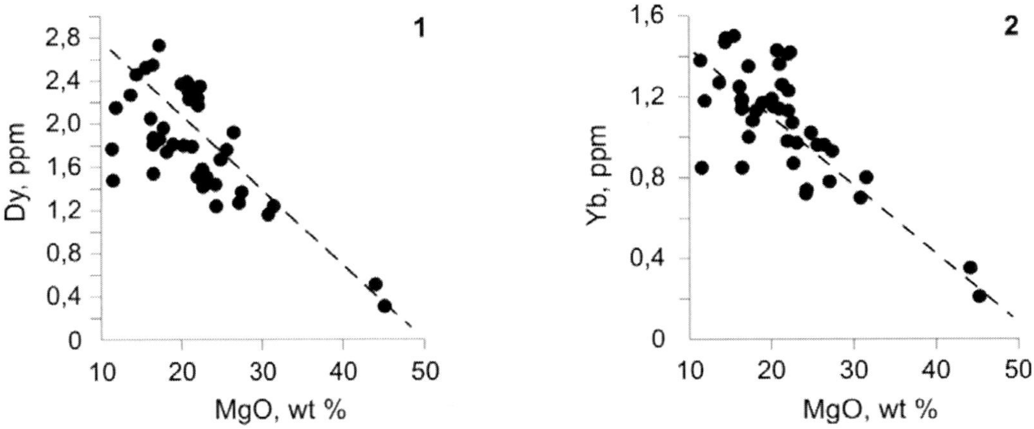

Figure 26. Graphs of correlation between Mg composition and Dy (1) and Yb (2) composition for komatiites from Gorgona (Columbia) [Revillon et al., 2000], Onwervaht [Jahn et at al., 1982], Barberton [Sun, Nesbit, 1978] (South Africa) provinces and Kambalda [Lesher, Arndt, 1995] and Perseverance [Barnes et al., 1995] provinces (Australia).

Bushveld Massif (South Africa)

In the area of this massif there are widely-spread fields of picrites as well as of high-magnesian and tholeiitic basalts, for which PGE, Au and La contents were determined [Davies, Tredoux, 1985]. The average La contents increase in the row from picrites (5.2 ppm) to magnesium basalts (24.8 ppm) and tholeiitic basalts (30.4 ppm). Average total Pd, Pt, Rh, Ru and Ir contents, on the contrary, decrease from picrites (45.2 ppb) to magnesium basalts (33.8 ppb) and tholeiitic basalts (19.0 ppb) (Table 13). The results of these analyses and their graphical interpretation (Figures 27, 28) allow us to conclude that between the contents of La, as of the other REE, and PGE contents in these volcanic rocks, there is an inverse relation. Moreover, as La contents are increasing in the same row of rocks the values of Pd/Ir parameter are also increasing (Figure 29). The observed correlations between REE and

PGE in volcanic rocks of this area, as it can be assumed, were the result of differences in the degree of partial melting of mantle sources during the generation of parent melts.

Table 13. La (ppm), and platinum group element (ppb) composition of picrites and high-Mg basalts from Bushveld pluton region (South Africa)

[Davies, Tredoux, 1985], NAA_										
Element	PEU1	PEU2	PEU3	Average	PES1	PEQ1	PEQ2	PEQ3	PEM1	EB13
Picrites					*Magnesian basalts*					
La	8,00	4,70	2,80	5,20	25,1	22,0	24,3	35,6	32,5	1,70
Ir	1,88	1,48	2,91	2,09	0,59	0,26	0,25	0,22	0,11	1,90
Ru	8,00	5,00	6,00	6,00	2,00	2,00	N.d.	2,00	N.d.	4,00
Rh	4,80	2,80	7,70	5,10	2,40	0,70	1,70	0,70	0,30	4,30
Pt	20,0	18,0	29,0	22,0	15,0	21,0	21,0	18,0	8,00	24,0
Pd	8,00	12,0	11,0	10,0	8,00	12,0	17,0	10,0	6,00	16,0
Total PGE	42,7	39,3	56,6	45,2	28,0	36,0	40,0	30,9	14,4	50,2
Pd/Ir	4,26	8,11	3,78	4,78	13,6	46,15	68,0	45,5	54,6	8,42
Pt/Ir	10,6	12,2	9,97	10,5	25,4	80,8	84,0	81,8	72,7	12,6

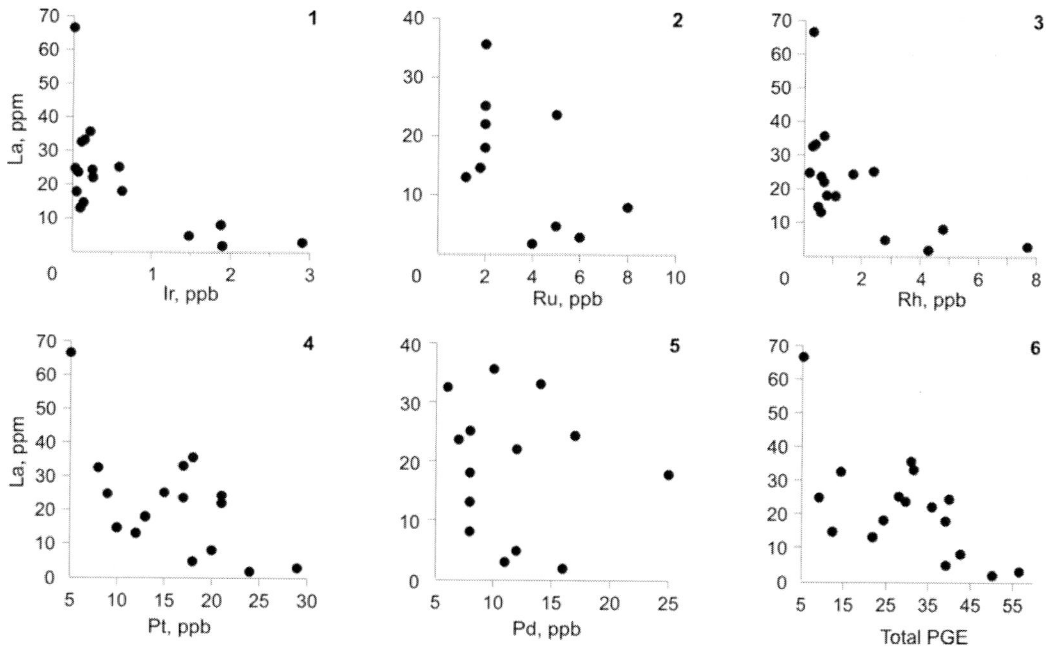

Figure 27. Graphs of correlation between La composition and Ir, Ru, Rh, Pt, Pd and total PGE composition for picrites, magnesium and tholeiite basalts from region of Bushveld pluton (South Africa) (data Table 13).

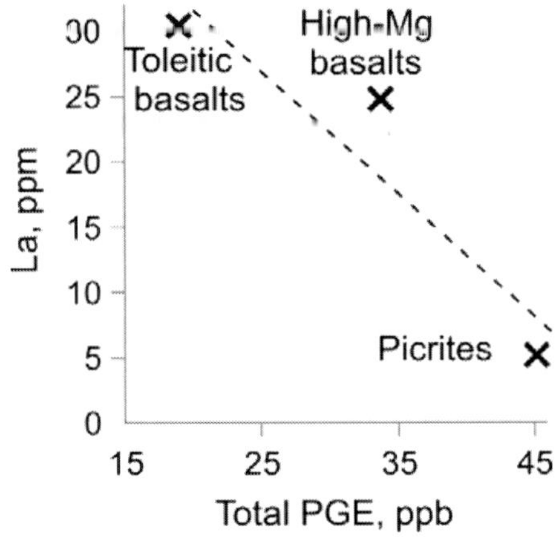

Figure 28. Graphs of correlation between La average composition and total PGE average composition for picrites, magnesium and tholeiite basalts from region of Bushveld pluton (South Africa) (data Table 13).

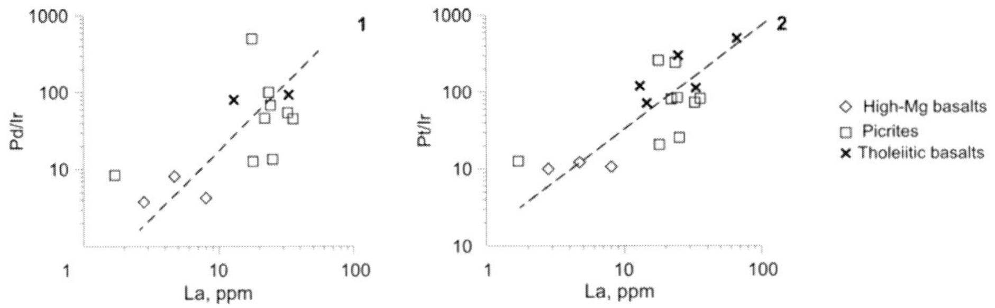

Figure 29. Graphs of correlation between La composition and Pd/Ir (1) and Pt/Ir (2) parameters values for picrites, magnesium and tholeiite basalts from region of Bushveld pluton (South Africa) (data Table 13).

Muskox Massif (Canada)

While studying this mafic-ultramafic massif located in the northwestern part of the Canadian Shield, there were data on the distribution of REE and PGE obtained for the rocks composing it: dunites, peridotites, websterites, olivine clinopyroxenites, gabbros, gabbro-norites and chromitites (Day et al., 2008). Almost all the rocks of this massif, including dunites, were relatively enriched with LREE (Table 14, Figure 30). It should be noted that by an increased level of REE accumulation and its distribution patterns the dunites of the Muskox massif are comparable to the samples that are dredged in the area of Romansh fault (Atlantic ocean) [Nikol'skaya & Kogarko, 1995]. At the same time, the rocks of the massif show a higher PGE concentration in comparison with rocks from most of the mafic-ultramafic massifs and complexes discussed above and the highest total PGE contents are characteristic for dunites and websterites. The chondrite-normalized contents of Os, Ir and Ru in all the rocks of the massif are inferior to the contents of Rh, Pt and Pd. In RGE distribution patterns of dunites, peridotites and websterites the maxima are observed for Pt and in the patterns of olivine clinopyroxenites and gabbros the maxima are observed for Pd. (Figure 31). The rocks of the massif, regardless of petrographic origin, slightly differ by the values of

Pd/Ir parameter that range in dunites – 0.28-85, in peridotites – 1.6-23, in websterites - 0.52-50, in clinopyroxenites and gabbros – 0.48-37. Variations of the values of Pd/Ir are due to both Pd dispersion and Ir dispersion.

Table 14. Rare earth element (ppm), platinum group element and rhenium (ppb) composition of rocks from Muskox massif (Canada)

	[Day et al., 2008], ICP-MS								
Element	N-75	N-115	N-129	S-44	S-51	S-60	S-102	S-110	S-128
	Dunites								
La	0,22	0,87	0,05	0,26	0,74	0,06	1,24	1,34	0,99
Ce	0,56	2,1	0,33	0,58	1,81	0,15	2,79	3,03	2,36
Pr	0,09	0,31	0,02	0,08	0,27	0,03	0,39	0,44	0,3
Nd	0,43	1,45	0,33	0,34	1,24	0,17	1,71	1,93	1,53
Sm	0,12	0,37	0,09	0,08	0,33	0,04	0,42	0,48	0,37
Eu	0,05	0,11	0,02	0,02	0,08	0,01	0,1	0,14	0,11
Gd	0,14	0,43	0,13	0,09	0,37	0,05	0,44	0,53	0,38
Tb	0,02	0,07	0,02	0,01	0,06	0,01	0,07	0,08	0,06
Dy	0,15	0,41	0,15	0,09	0,39	0,05	0,41	0,48	0,36
Ho	0,03	0,08	0,03	0,02	0,08	0,01	0,08	0,09	0,07
Er	0,11	0,22	0,1	0,06	0,21	0,04	0,21	0,25	0,19
Tm	0,02	0,03	0,02	0,01	0,03	0,01	0,03	0,04	0,03
Yb	0,15	0,22	0,12	0,07	0,2	0,07	0,21	0,24	0,18
Lu	0,03	0,03	0,02	0,01	0,03	0,01	0,04	0,04	0,03
Total REE	2,12	6,70	1,43	1,72	5,84	0,71	8,14	9,11	6,96
(La/Yb)*n*	0,99	2,67	0,28	2,51	2,50	0,58	3,99	3,77	3,71
Os	77,4	1,62	19,3	7,74	5,71	0,956	2,04	3,89	6,25
Ir	50,2	1,49	11,4	9,77	3,04	0,575	1,31	2,93	5,50
Ru	49,7	8,14	78,9	17,2	9,11	2,16	11,0	7,60	15,6
Pt	36,9	25,6	48,1	8,60	38,0	99,8	17,5	29,1	115
Pd	21,6	37,6	201	2,67	26,4	80,9	14,6	28,5	20,6
Total PGE	236	74	358	46	82	184	46	72	163
Re	0,312	0,509	2,70	0,022	0,105	0,035	0,103	0,079	0,105
Pd/Ir	0,43	25,2	17,6	0,27	8,70	141	11,2	9,73	3,75
Pt/Ir	0,73	17,2	4,24	0,88	12,5	174	13,4	9,95	20,9

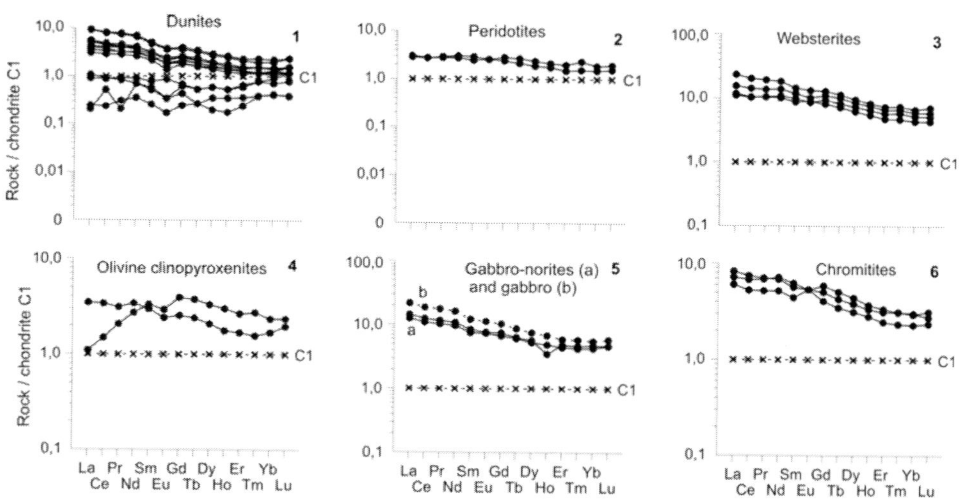

Figure 30. Chondrite-normalized REE patterns for rocks from Muskox massif (Canada) (data Table 14).

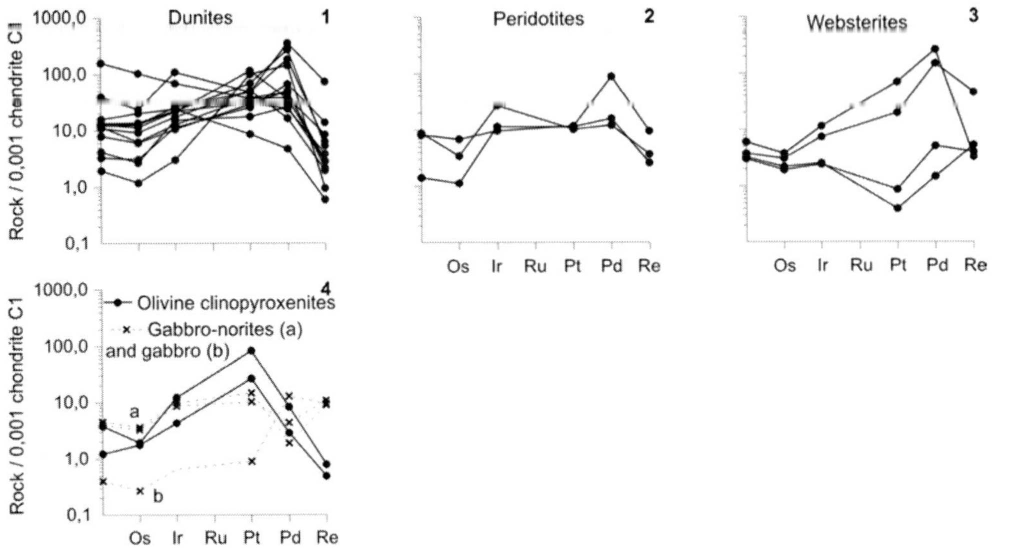

Figure 31. Chondrite-normalized PGE patterns for rocks from Muskox massif (Canada) (data Table 14).

The following diagrams show the correlations between the contents of REE and PGE in rocks of Muskox massif (Figure 32). Judging by the location of figurative points in them, we can assume quite certainly that the rocks of the massif have a more or less distinctly marked inverse relation between contents of elements from these groups. In turn, the location of points in the diagrams illustrating the correlation between the contents of Yb and the values of Pd/Ir and Pt/Ir parameters shows that at comparatively limited variations of REE contents the values of Pd/Ir and Pt/Ir vary in rather wide ranges, which is most clearly illustrated by the example of points corresponding to the dunites composition (Figure 33).

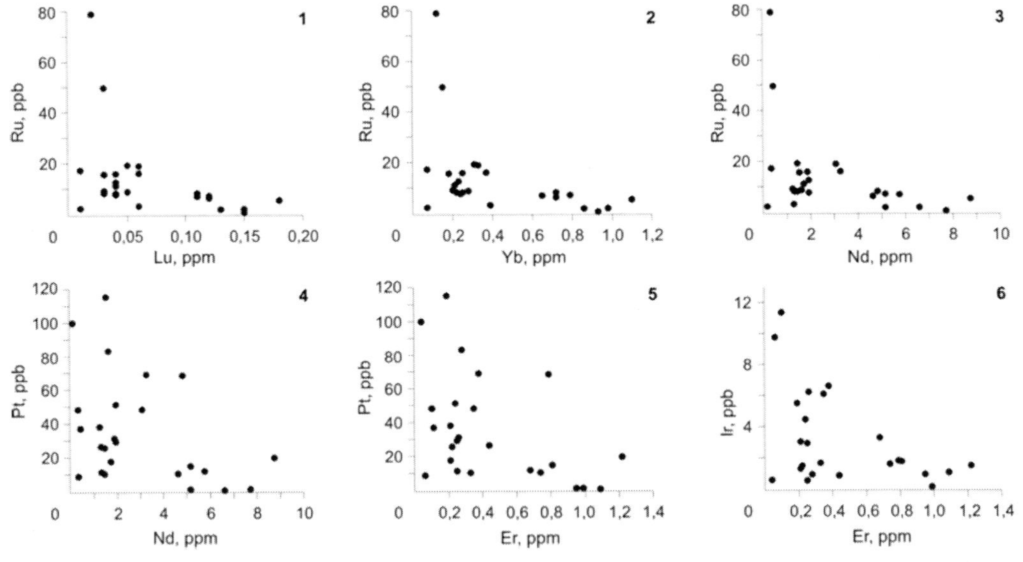

Figure 32. Graphs of correlation between some REE composition and PGE composition for rocks from Muskox massif (Canada) (data Table 14).

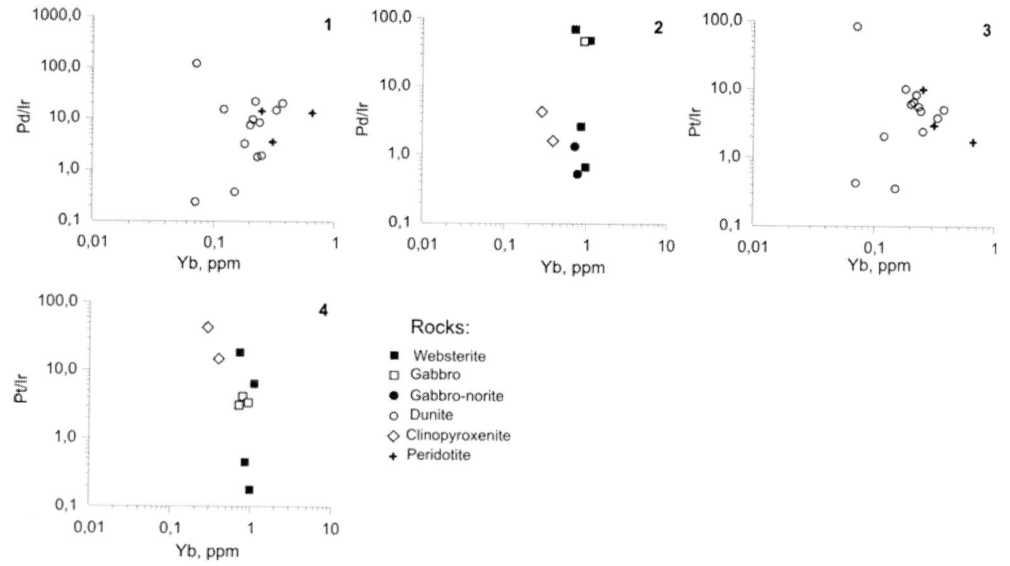

Figure 33. Graphs of correlation between Yb composition and Pd/Ir and Pt/Ir parameters values for rocks from Muskox massif (Canada) (data Table 14).

Galmoenan massif (Koryak highland, Russia) is located in the Koryak highland (Kamchatka). Detailed petrological and geochemical studies on the massif were performed in the course of exploration and exploitation of closely located and genetically associated large placer deposit of platinum group metals [Koryak-Kamchatka region ..., 2002].

Body of the massif elongated in the submeridional direction, has a length of about 16 km and a width of 2-4 km, and is exposed on an area of approximately 48 sq. km. It is composed of dunites that make up its interior part, less commonly of olivine, plagioclase-bearing and magnetite clinopyroxenites that compose a 200-700 m wide zone surrounding the dunite body, while among the pyroxenites some xenogeny isolations of wehrlites and serpentinites occur.

Along the contacts of pyroxenite zone with terrigenic-volcanic formations that are framing the massif there is a gabbro intrusion fragmentarily uncovered by erosion.

Based on the results of comprehensive research, including numerical modeling, Batanova et al. [2011] concluded that the diversity of ultramafic and mafic rocks that form the massif arose from an intrachamber differentiation of single upper mantle mafic melt.

However, our observations and, in particular, the presence of xenogeny isolations of wehrlites and serpentinites among pyroxenites do not allow to exclude the possibility of existence of an alternative model of the genesis of this massif, which is based on an idea of polygenic formation of complex mafic-ultramafic massifs [Lesnov, 1981, 1982, 1986].

According to the model, the dunite body of the considered massif is an earlier protrusive block of upper mantle restites. In its peripheral part the dunites were transformed into clinopyroxenites as a result of exposure to later mantle mafic melts and their fluids.

Mafic melts intruded along tectonic contacts of dunite protrusion with the framing rocks and then formed gabbro intrusion that is unevenly uncovered on the perimeter of this complex massif. However, the applicability of this alternative model of formation to this mafic-ultramafic massif requires further studies.

Table 15. Rare earth and platinum group elements, gold, and rhenium (ppm) composition of dunites from Gal'moenansky massif (Koryak highland, Russia)

[Koryakia Kamchatka region…, 2002], LA ICP MS.

Element	Gal-2	Gal -7	Gal -9	Gal -11	Gal -12	Gal -13	Gal -15	Gal -16	Gal -17	Gal-18	Gal-19
La	0,099	0,02	0,035	0,02	0,02	0,02	0,324	0,099	0,035	0,098	0,113
Ce	0,132	0,02	0,166	0,02	0,021	0,02	0,406	0,187	0,147	0,177	0,137
Pr	0,011	0,01	0,01	0,01	0,01	0,01	0,033	0,013	0,01	0,017	0,01
Nd	0,034	0,02	0,022	0,044	0,015	0,02	0,02	0,048	0,029	0,128	0,02
Sm	0,02	0,02	0,02	0,02	0,02	0,032	0,02	0,02	0,031	0,033	0,015
Eu	0,02	0,02	0,02	0,02	0,02	0,02	0,02	0,02	0,02	0,02	0,02
Gd	0,02	0,02	0,02	0,02	0,02	0,02	0,02	0,02	0,02	0,029	0,017
Tb	0,01	0,01	0,01	0,01	0,01	0,01	0,01	0,01	0,01	0,01	0,01
Dy	0,02	0,02	0,02	0,015	0,02	0,02	0,02	0,02	0,019	0,02	0,036
Ho	0,01	0,01	0,01	0,01	0,01	0,01	0,01	0,01	0,01	0,01	0,01
Er	0,02	0,02	0,015	0,018	0,02	0,016	0,015	0,02	0,014	0,017	0,038
Tm	0,01	0,01	0,01	0,01	0,01	0,01	0,01	0,01	0,01	0,01	0,01
Yb	0,02	0,027	0,029	0,02	0,02	0,02	0,02	0,02	0,019	0,014	0,04
Lu	0,01	0,01	0,01	0,01	0,01	0,01	0,01	0,01	0,01	0,01	0,01
Total REE	0,44	0,24	0,40	0,25	0,23	0,24	0,94	0,51	0,38	0,59	0,49
Os	0,02	0,038	0,055	0,02	0,024	0,02	0,02	0,022	0,02	0,02	0,02
Ir	0,01	0,01	0,01	0,015	0,01	0,082	0,01	0,01	0,01	0,01	0,01
Ru	0,037	0,02	0,02	0,02	0,025	0,02	0,02	0,02	0,02	0,019	0,02
Rh	0,013	0,016	0,1	0,034	0,01	0,05	0,01	0,01	0,01	0,01	0,01
Pt	0,035	0,367	0,115	0,291	0,097	0,492	0,022	0,083	0,02	0,11	0,021
Pd	0,11	0,489	0,311	0,093	0,17	0,02	0,034	0,09	0,043	0,087	0,02
Total PGE	0,23	0,94	0,61	0,47	0,34	0,68	0,12	0,24	0,12	0,26	0,10
Re	1,497	0,01	0,01	0,01	0,01	0,01	0,045	0,01	0,01	0,496	0,01
Au	0,02	0,02	0,047	0,02	0,063	0,02	0,02	0,02	0,02	0,02	0,02
Element	**Gal -20**	**Gal -21**	**Gal -22**	**Gal -24**	**Gal -25**	**Gal -26**	**Gal -27**	**Gal -28**	**Gal -29**	**Gal -30**	
La	0,02	0,02	0,02	0,06	0,02	0,03	0,14	0,11	0,03	0,04	
Ce	0,02	0,02	0,066	0,06	0,02	0,04	0,17	0,2	0,05	0,15	
Pr	0,01	0,01	0,01	0,01	0,01	0,01	0,02	0,07	0,01	0,07	
Nd	0,02	0,02	0,085	0,02	0,02	0,02	0,04	0,36	0,11	0,36	
Sm	0,021	0,02	0,012	0,02	0,02	0,02	0,03	0,13	0,03	0,15	
Eu	0,02	0,02	0,02	0,02	0,02	0,02	0,02	0,06	0,02	0,06	
Gd	0,02	0,02	0,02	0,02	0,02	0,03	0,05	0,016	0,04	0,16	
Tb	0,01	0,01	0,01	0,01	0,01	0,01	0,01	0,03	0,01	0,03	
Dy	0,02	0,02	0,02	0,02	0,02	0,02	0,02	0,12	0,05	0,14	
Ho	0,01	0,01	0,01	0,01	0,01	0,01	0,01	0,05	0,02	0,03	
Er	0,021	0,011	0,01	0,02	0,02	0,02	0,04	0,11	0,03	0,08	
Tm	0,01	0,01	0,01	0,01	0,01	0,01	0,01	0,02	0,01	0,02	
Yb	0,02	0,018	0,021	0,02	0,02	0,03	0,03	0,09	0,04	0,07	
Lu	0,01	0,01	0,01	0,01	0,01	0,01	0,01	0,02	0,01	0,01	
Total REE	0,23	0,22	0,32	0,31	0,23	0,28	0,60	1,39	0,46	1,37	
Os	0,02	0,028	0,02	0,02	0,02	0,02	0,02	0,02	0,02	0,02	
Ir	0,01	0,01	0,01	0,01	0,01	0,01	0,01	0,01	0,01	0,01	
Ru	0,02	0,02	0,02	0,02	0,02	0,02	0,02	0,02	0,02	0,02	
Rh	0,011	0,01	0,01	0,01	0,01	0,01	0,01	0,01	0,01	0,01	
Pt	0,21	0,106	0,062	0,32	0,04	0,2	0,02	0,06	0,02	0,02	
Pd	0,02	0,02	0,02	0,02	0,02	0,02	0,02	0,02	0,02	0,02	
Total PGE	0,29	0,19	0,14	0,40	0,12	0,28	0,10	0,14	0,10	0,10	
Re	0,01	0,01	0,01	0,03	0,05	0,16	0,17	0,04	0,01	0,15	
Au	0,02	0,02	0,027	0,02	0,02	0,02	0,02	0,02	0,02	0,02	

Published data on the contents of REE and PGE in dunites from Galmoenan massif (Table 15), as well as diagrams based on them (Figure 34) give ground to assume that in this case there is an inverse relation between the contents of the elements of these two groups.

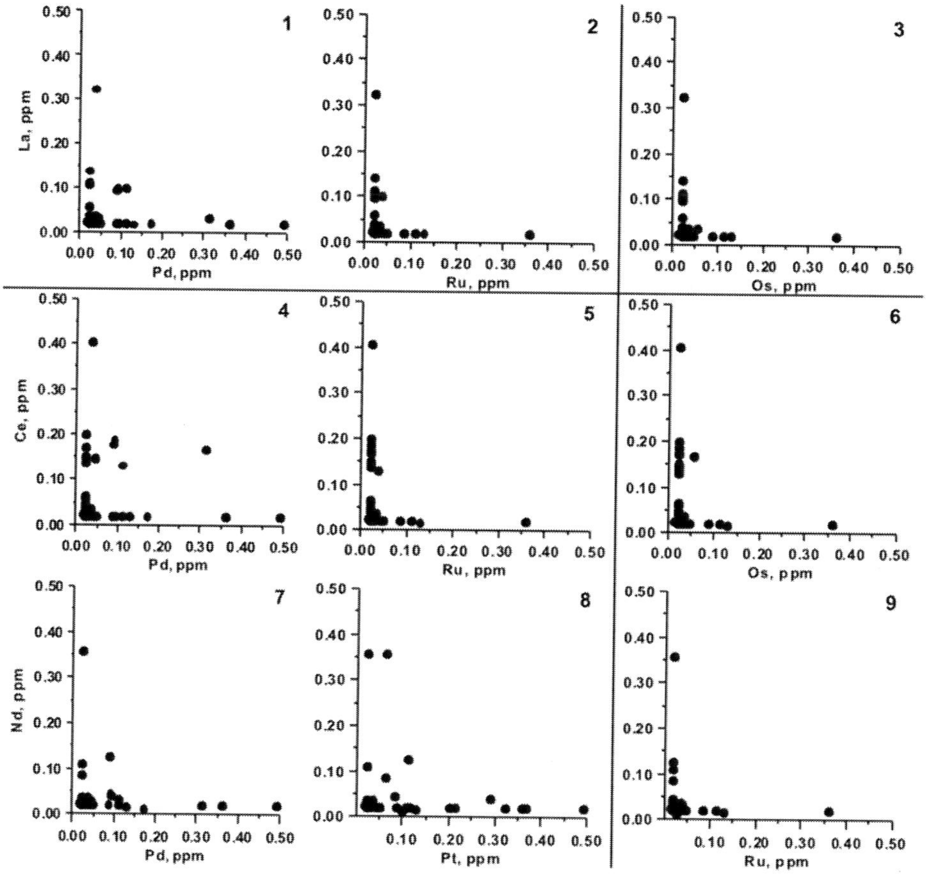

Figure 34. Graphs of correlation between REE composition and PGE composition for dunites from Gal'moenansky massif (Koryak highland, Russia) (data Table 15).

Discussion and Conclusion

Based on the synthesis of available analytical data on PGE contents in different types of ultramafic and mafic rocks Barnes et al. [1985] proposed to single out three following mechanisms of fractionation of these elements, which were implemented at different magmatic processes: 1) partial melting of mantle sources, 2) crystallization fractionation of magmatic melts, and 3) reallocation at late-magmatic and post-magmatic transformations of rocks. These researchers believed that PGE can not be in the form of isomorphous impurity in the structure of minerals such as olivine and chromite.

A similar point of view was later expressed by Borisov [2005], who believed that all PGE are incompatible with the structure of major silicate phases composing the upper mantle -

olivine, orthopyroxene, clinopyroxene and garnet. According to Borisov, the values of K_d (silicate mineral / melt) for all PGE are less than 1. He also suggested that, when the degree of partial melting of mantle sources is 20%, an almost full transition into melt of all those phases that originally contained PGE occurs.

Calculations made by McDonough & S.-S. Sun [1995] showed that the average PGE contents in non-depleted upper mantle are (ppb): Os – 3.4; Ir – 3.2; Ru - 5; Pt – 7.1; Pd – 3.9. Similar estimates for primitive mantle were previously given by Rollinson [1993] (ppb): Os – 3.3; Ir – 3.6; Ru – 4.3; Pt - 7. W. McDonough & S.-S. Sun noted that in basaltic smelting formed during partial melting of non-depleted upper mantle the overall balance of PGE should grow by about half-order compared to the original mantle source, thus to be (ppb): Os - 17; Ir - 16; Ru - 25; Pt - 35; Pd - 20. When extrapolating these values to a lower degree of partial melting of the mantle source, for example, to 10%, the PGE contents in basaltic smelting, obviously, will be (ppb): Os – 8.5; Ir - 8; Ru – 12.5; Pt – 17.5; Pd - 10.

To add more, Barnes et al. [1985] proposed to divide the PGE group into two subgroups 1) subgroup of iridium (Os, Ir, Ru), and 2) subgroup of palladium (Rh, Pt, Pd); however, as they emphasized, such division should not be considered as an analogy for the separation of REE group into subgroups of LREE and HREE.

These researchers also pointed at the possibility of using Pd/Ir parameter as one of the geochemical indicators for discrimination and classification of rocks of ultramafic and mafic composition. They estimated the average values of this parameter and found it increasing from peridotite komatiites (10) to pyroxenite komatiites (30) and further to the basalts of continents and ocean floor (100). Note that in plutonic varieties of ultramafic and mafic rocks the values of this parameter vary over a wider range, for example, in some varieties of samples from mafic-ultramafic Stillwater massif (Montana, USA) the value of Pd/Ir reached 865, while in ore chromitites of ophiolite associations it is close to 0.1.

Taking into account the data mentioned above on the distribution of REE and PGE in ultramafic and mafic rocks, as well as on correlations between the contents of the elements of these two groups, one can assume that in those mantle basaltic smelting that is formed when the degree of partial melting was 20%, the concentration of REE, which are elements that are relatively easy to transform into melt, should be lower than their concentration in basaltic smelting, which arose at about 10% degree of partial melting of a similar in composition mantle source. It is obvious that the increase of degree of partial melting of mantle source was accompanied by a decrease in REE contents of each subsequent portion of melt and simultaneous relative enrichment of this portion with platinum group elements.

However, as can be seen from the analytical data mentioned above in this chapter, even at the highest degrees of partial melting and resulting intense depletion of the mantle source by fusible components the ultramafic restites formed as a result of that process were not fully sterilized with respect to REE.

Summarizing the considered materials on REE and PGE relations in different types of rocks and mafic-ultramafic associations, we can conclude that partial melting of non-depleted upper mantle, resulted in the formation of ultramafic restites and basaltic melts complementary to them, in fact was the earliest stage of a multistage fractionation and redistribution of PGE and REE within the mantle-crust system.

At the same time with increasing of degrees of partial melting of mantle sources in the forming ultramafic restites, as well as in complementary mafic smelting, and then - also in the products of their crystallization, an increase in PGE concentrations was accompanied by a decrease in REE concentrations. As a result, in these rocks we can observe an inverse relation between the contents of the elements from these two groups contrasting in their properties.

Apparently, the same type of mechanism could provide a described by some specific examples "phenomenon" of geochemical antagonism between REE and PGE. However, such interpretation of the relations between these elements in ultramafic and mafic rocks can be considered only as a first approximation to the forthcoming solution of these complex petrological and geochemical problems.

References

Anders, E., Grevesse, N. (1989) Abundances of the elements: meteoritic and solar. *Geochim. Cosmochim. Acta,* 53, 197–214.

Balykin, P.A., Polyakov, G.V., Hanski E. et al. (2004) The Late Permian komatiite-basalt complex in the Song Da Rift, North-Western Vietnam. *J. Geology. Series B*, 23, 52–64.

Barnes, S.-J., Naldrett, A.J., Gorton, M.P. (1985) The origin of fractionation of platinum-group elements in terrestrial magmas. *Chemical Geology*, 53, 303–323.

Batanova, V.G., Pertsev, A.N., Kamenetsky, V.S., Mochalov, A.G., Sobolev, A.V. (2011) Crustal evolution of island-arc ultramafic magma: Galmoenan pyroxenite-dunite plutonic complex, Koryak highland (Far East Russia). *J. Petrolojy*, 46 (7), 1345-1366.

Borisov, A.A. (2005). Crystallization and stability of alloys of precious metals in magmatic process. *Geology of ore deposits,* 47 (6), 516–523 (in Rus.).

Borodina, E.V., Egorova, V.V., Izokh A.E. (2004). Petrology of Ordovician collision a peridotite-gabbro of files. *Geologia i geofizika*, 45 (9), 1074–1091 (in Rus.).

Brugmann, G.E., Arndt, N.T., Hofmann, A.W., Tobschall, H.J. (1987) Noble metal abundances in komatiite suites from Alexo, Ontario and Gorgona Island, Columbia. *Geochim. Cosmochim. Acta,* 51, 2159–2169.

Davies, G., Tredoux, M. (1985) The platinum-group element and gold contents of the marginal rocks and sills of Bushveld complex. *Economic Geology*, 80, 838–848.

Day et al., 2008;

Day J.M.D., Pearson D.G., Hulbert L.J. (2008) Rhenium-osmium isotope and platinum-group element constraints on the origin and evolution ofepy 1,27 Ga Muskox layered intrusion. *J. Petrol.,* 49 (7), 1255–1295.

Dmitrenko, G.G., Lesnov, F.P., Makhorkina, T.I. et al. (1991) The platinum minerals in chromitites from Naransky massif (Western Mongolia). *Doklady of Academy of Sciences of USSR*, 317 (5), 1220–1222 (in Rus.).

Edwards, R.L., Wasserburg, G.J. (1985) The age and emplacement of obducted oceanic crust in the Urals from Sm-Nd and Rb-Sr systematics. *Earth Planet. Sci. Lett.,* 72, 389–404.

Evensen, N.M., Hamilton, P.J., O'Nions, R.K. (1978) Rare earth abundances in chondritic meteorites. *Geochim. Cosmochim. Acta.,* 42, 1199–1212.

Fershtater, G.B. (2004) On the nature of the Silurian-Early Devonian mafic-ultramafic intrusions associated with the ophiolite of the Southern Urals. *Lithosphere,* 4, 3-29 (in Rus.).

Fershtater, G.B., Bea, F. (1996) Geochemical typing of Uralian ophiolite. *Geochemiya*, 3, 195-218 (in Rus.).

Frey, F.A., Haskin, L.A., Haskin, M.A. (1971) Rare earth abundances in some ultramafic rocks. *J. Geophys. Res.,* 76 (8), 2057–2070.

Gladney, E.S., Jones, E.A., Nickell, E.J., Roelandts, I. (1991) 1988 Compilation of Elemental Concentration Data for USGS DTS-1, G-1, PCC-1 and W-1. *Geostandards Newsletter,* 15, 199–396.

Hanski, E., Walker R.J., Huhma H. et al. (2004). Origin of Permian-Triassic komatiites, northwestern Vietnam. *Contrib. Mineral. Petrol.,* 147, 453–469.

Jahn, B., Gruau, G., Glikson, A.Y. (1982) Komatiites of the Onverwacht group, South Africa: REE geochemistry, Sm/Nd age and mantle evolution. *Contrib. Mineral. Petrol.,* 80, 25–40.

Karipi, S., Tsikouras, B., Hatzipanagiotou, K. (2006) The petrogenesis and tectonic setting of ultramafic rocks from Iti and Kallidromon mountains, continental central Greece: vestiges of the Pindos ocean. *Canadian Mineralogist,* 44 (Part 1), 267–287.

Koryak-Kamchatka region – New platinumferous province of Russia (2002) (Sci. Eds. V.P. Zaytsev, A.F. Litvinov, E.A. Landa). Sanct-Peterburg. Publishing House "Kartfabrika VSEGEI. 383 p. (in Rus.).

Lazarenkov, V.G., Balmasova, E.A., Vaganov, P.A., Onischina, N.M. (1989) The distribution of precious metals and rare earth elements in the ore chromitite from Burakovo-Aganozersky massif. *Newsletter of Leningradsky State University. Series 7, Issue 4*, 68-70 (in Rus.).

Lesher, C.M., Arndt N.T. (1995) REE and Nd isotope geochemistry, petrogenesis and volcanic evolution of contaminated komatiites at Kambalda, Western Australia. *Lithos,* 34 (3–4), 127–157.

Lesnov, F.P. (1979) On the mafic-ultramafic associations of Mongolia. *Geology and Magmatism of Mongolia.* Moscow: Publishing House Nauka, 156–157 (in Rus.).

Lesnov, F.P. (1981) Structural and genetic relationships ultramafic rocks and gabbroid in the ophiolite belts of Mongolia. *Questions of magmatism and metallogeny of the Mongolian People's Republic*. Novosibirsk: Publishing House Nauka, Siberian branch, 62–71 (in Rus.).Lesnov, F.P. (1982) Naransky polygenic mafic-ultramafic pluton (Western Mongolia). *Ultramafic association of folded regions. Issue 1.* Novosibirsk: Publishing House of Institute geology and geophysics of Academy of Sciences of USSR, 58–95 (in Rus.).Lesnov, F.P. (1986) *Petrochemistry of polygenic mafic-ultramafic plutons of folded regions.* Novosibirsk: Publishing House Nauka. Siberian branch, 136 p. (in Rus.).

Lesnov, F.P. (1993) Regularities in the distribution of platinum group elements in the Naran polygenic mafic-ultramafic pluton (Western Mongolia). *Ultramafic association of folded regions. Issue 7.* Novosibirsk: Publishing House of Institute geology and geophysics of Academy of Sciences of USSR, 6-35 (in Rus.).

Lesnov, F.P. (2003a) Role of rare earth and platinum group elements in the problem of typing ultramafic-mafic formations. *Modern Problems formation analysis, petrology and igneous rocks of the ore.* Novosibirsk: Publishing House "Geo", Siberian branch, 195-196 (in Rus.).

Lesnov, F.P. (2003b) The correlation ofplatinum and rare earth elements in mafic-ultramafic rocks as a possible criterion for their typing. *Geochemistry of igneous rocks. Proceedings of the 21-th All-Russian Seminar-School "Alkaline Magmatism of the Earth.* Apatity: Publishing House of Geological Institute of Kola Science Centre of the Russian Academy of Sciences, 101-102 (in Rus.).

Lesnov, F.P. (2006) The correlation between rare earth and platinum group elements as reflection of partial melting process of mantle resource at forming of mafic and ultramafic rocks. *Ophiolite: geology, petrology, metallogeny, geodynamics. Proceedings of the International Scientific Conference.* Yekaterinburg. Publishing House of Institute geology and geochemistry. Ural branch of Russian Academy of Sciences, 209-213 (in Rus.).

Lesnov, F.P. (2009) Contrast fractionation of rare earth and platinum group elements at partial melting and depletion of upper mantle. *Geology and mineral resources of European North-East of Russia. Vol 2.* Syktyvkar: Publishing House of Institute geology. Komi Science Center. Ural brench of Russian Academy of Sciences, 345-347 (in Rus.).

Lesnov, F.P. (2010) Rare Earth Elements in ultramafic and mafic rocks and their minerals. Main types rocks. *Rock-forming minerals.* Leiden: Publishing House CRC. Taylor & Francis Group, 580 p.

Lesnov, F.P., Balykin, P.A. (2005) On the additional evidences of "antagonism" of platinum and rare earth elements in igneous rocks. *Precious and Rare Metals of Siberia and the Far East.* Irkutsk: Publishing House of Institute of geochemistry of Siberian branch of Russian Academy of Sciences, 17-19 (in Rus.).

Lesnov, F.P., Balykin, P.A. (2007) Distribution of rare earth and platinum group elements in rocks of Kokpektinsky dunite-troctolite-gabbro massif (Southern Urals). *Ultrabasite-mafic complexes of folded regions: Proceedings of the International conference.* Irkutsk: Publishing house of Irkutsk State technical University, 192-196 (in Rus.).

Lesnov, F.P., Oydup, Ch.K. (2002) On the "antagonism" of palladium and platinum with rare earth elements in rocks of ultramafic-mafic massifs. *Geology, genesis and questions of development of complex deposits of precious metal. Proceedings of the All-Russian Symposium.* Moscow: Publishing House of IGEM of Russian Academy of Sciences, 48–52 (in Rus.).

Lesnov, F.P., Simonov, V.A. (1989). Chromitites of Naransky mafic-ultramafic massif in Western Mongolia (chemical composition and gas components). *Ultramafic association of folded regions. Issue 5.* Novosibirsk: Publishing House of Institute of geology and geophysics of Siberian branch of Academy of Sciences of USSR, 65-85 (in Rus.).

Lesnov, F.P., Tsimbalist, V.G. (1983) Distribution of platinum group elements and gold in mafic-ultramafic associations of Mongolia. *Ultramafic association of folded regions. Issue 2.* Novosibirsk: Publishing House of Institute of geology and geophysics of Siberian branch of Academy of Sciences of USSR, 66-71 (in Rus.).

Lesnov, F.P., Mongush, A.A., Tsimbalist, V.G. (2000) On some features of the geochemistry of platinum group elements in rocks from massifs of mazhalyksky mafic-ultramafic complex (Tuva Republic). *Geology and mineral resources of Central Siberia.* Krasnoyarsk: Publishing House of KNIIGiMS, 164-166 (in Rus.).

Lesnov, F.P., Mongush, A.A., Oydup, Ch.K. (2001). Geochemical communication of platinum elements and rare earth elements in mafic-ultramafic rocks of Tuva (first data). *Platinum in geological formations of Siberia.* Abstracts. All-Russian Seminar. Krasnoyarsk: Publishing House of KNIIGiMS, 107-110 (in Rus.).

Lesnov, F.P., Mongush, A.A., Mel’gunov, M.S. (2004) On some features of the geochemistry of platinum group elements and rare earth elements in rocks from the ultramafic-mafic massifs (Tuva). *The state and exploration of natural resources of Tuva and adjacent regions of Central Asia. Geoecology of the environment and society.* Kyzyl: Publishing

House of Tuvinian Institute for exploration of natural resources of Siberian branch of the Russian Academy of Sciences, 54-60 (in Rus.).

Lesnov, F.P., Mongush, A.A., Anoshin, G.N. et al. (2005a) Investigating the distribution of rare earth and platinum group elements in rocks from the mafic-ultramafic massifs of Tuva by ICP-MS method (first data). *Petrology of igneous and metamorphic complexes.* Tomsk: Publishing House of Tomsk State University, 268-277 (in Rus.).

Lesnov, F.P., Mongush, A.A., Palessky, S.V. (2005b) The correlation between platinum and rare earth elements in some rocks, mafic-ultramafic massifs of Tuva (according to ICP-MS method). *Precious and Rare Metals of Siberia and the Far East: the ore-forming system of integrated and non-traditional fields of ore types.* Irkutsk: Publishing House of Institute of geochemistry of Siberian branch of the Russian Academy of Sciences, 19-22 (in Rus.).

McDonough W.F., Sun S.-S. (1995) The composition of the Earth. *Chem. Geol.,* 120, 223–253.

Melcher F., Grun W., Thalhammer T.V., Thalhammer O.A.R. (1999) The giant chromites deposits at Kempirsai, Urals: constraints from trace element (PGE, REE) and isotope data. *Mineralium Deposita,* 34, 250–272.

Mongush, A.A., Oydup, Ch.K., Agafonov, L.V., Lesnov, F.P. (1999) Geological position and platinum-bearing of Lower Cambrian mafic-ultramafic massifs of mazhalyksky complex (Tuva). *Geology and Mineral Resources of the Krasnoyarsk Territory.* Krasnoyarsk: Publishing House of KNIIGiMS, 183-188 (in Rus.).

Mongush, A.A,. Agafonov, L.V., Tunay, G.O. (2005) Mineralogy of the upper part of the fiber Bulkinsky gabbroic massif (Western Sayan). *Ultramafic-mafic complexes of folded regions of the Precambrian.* Ulan-Ude: Publishing House of Buriat Science Centre of of Siberian branch of the Russian Academy of Sciences, 62-64 (in Rus.).

Naldrett, A.J., Fedorenko, V.A., Latfut, P.K. et al. (1994) Deposits of Ni-Cu-PGE of Norilsk area, Siberia: their formation in the feeding channels of plateau basaltic volcanism. *7-th International Platinum Symposium: Abstracts.* Moscow: Publishing House Moscow contact, 78 (in Rus.).

Nikol'skaya, N.E., Kogarko, L.N. (1995) Geochemical features of ultramafic rocks from Romanche fault (Equatorial Atlantic). *Geokhimiya,* 9, 1280-1295 (in Rus.).

Oydup, Ch.K., Lesnov, F.P., Palessky, S.V. et al. (2007) Distribution of rare earth, platinum and other trace elements in rocks from the mafic-ultramafic massifs of ophiolite association in South-Western Tuva (according to the method of ICP-MS). *Status and development of natural resources of Tuva and adjacent regions of Central Asia. Geoecology of the environment and society. Vol. 9.* Kyzyl: Publishing House of Tuvinian Institute for exploration of natural resources of Siberian branch of the Russian Academy of Sciences, 116-128 (in Rus.).

Palessky, S.V. (2008) Determination of rare and trace elements by mass-spectrometry inductive coupled plasma method. *Abstract of Dissertation. PhD. Chemical sciences.* Novosibirsk: Publishing House of Institute of Inorganic Chemistry of Siberian branch of Russian Academy of Sciences, 18 p. (in Rus.).

Pallister, J.S., Knight R.J. (1981) Rare-earth element geochemistry of the Samail ophiolite near Ibrea, Oman. *J. Geoph. Res.,* 86 (B4), 2673–2697.

Pearson D.G., Woodland S.J. (2000) Solvent extraction/anion exchange separation and determination of PGEs (Os, Ir, Pt, Pd, Ru) and Re-Os isotopes in geological samples by isotope dilution ICP-MS. *Chem. Geol.*, 165, 87–107.

Petrology of postharzburgite intrusives of Kempirsay-Khabarninsky ophiolite association (South Urals) // P. Balykin at al. Sverdlovsk: Publishing House IGG, Ural Branch of AS USSR. 1991. 160 p (in Rus.).

Pinus, G.V., Velinskii, V.V., Lesnov, F.P. (1976) Ultramafic belts of Central Asia, and some general questions of petrology of ultramafic rocks. *Problems of petrology*. Moscow: Publishing House Nauka, 94-104 (in Rus.).

Pinus, G.V., Agafonov, L.V., Lesnov, F.P. (1984) *Alpine ultramafic rocks of Mongolia.* Moscow: Publishing House Nauka, 200 p. (in Rus.).

Poliakov, G.V., Bognibov, V.I. (1979) Early Paleozoic peridotite-pyroxenite-gabbro-norite complex of Salairide of Eastern Tuva. *Mafic and ultramafic complexes of Siberia.* Novosibirsk: Siberian branch of Publishing House Nauka, 118-126 (in Rus).

Poliakov, G.V., Bognibov, V.I., Izokh, A.E. et al. (1984) Peridotite-pyroxenite-gabbro-norite formation of East Tuva and North-Western Mongolia. *Plutonic formations of Tuva and their ore-bearing.* Novosibirsk: Siberian branch of Publishing House Nauka, 3-57 (in Rus.).

Rivalenti, G., Mazzucchelli, M., Vannucci, R. et al. (1995) The relationship between websterite and peridotite in the Balmuccia peridotite massif (NW Italy) as revealed by trace element variations in clinopyroxene. *Contrib. Mineral. Petrol.*, 121 (3), 275–288.

Rollinson, H.R. (1993) *Using Geochemical Data: Evaluation, Presentation, Interpretation*. Publishing House Longman Group, UK, 152 p.

Roman'ko, A., Kuznetsov, G., Savicheva, A. et al. (1994) Early Proterozoic magmatism in the Eastern Baltic shield and metallogeny of the peculiarities of the PGE. *7-th International Platinum Symposium: Abstracts.* Moscow: Publishing House Moscow Contact, 78 (in Rus.).

Rudnev S.N., Vladimirov A.G., Ponomarchuk V.A. et al.(2006) *Kaakhemsky polychronal batholith (West Tuva): composition, age, sources and geodynamic position* // Litosfera. (2), 3-33.

Sal'nikova, E.B., Kovach, V., Kozakov, I.K. et al. (2004) Age and geodynamic position of peridotite-pyroxenite-gabbro-anartosite mazhalyksky complex, Eastern Tuva. *Petrologiya,* 12 (6), 656-662 (in Rus.).

Schmidt, G., Witt-Eickschen, G., Palme, H. et al. (2003) Highly iderophile elements (PGE, Re and Au) in mantle xenoliths from the West Eifel volcanic field (Germany). *Chem. Geol.,* 196, 77–105.

Sharma, M., Wasserburg, G.J. (1996) The neodymium isotopic compositions and rare earth patterns in highly depleted ultramafic rocks. *Geochim. Cosmochim. Act,* 60, 4537–4550.

Sharma, M., Wasserburg, G.J., Papanastassiou, D.A. et al. (1995) High $^{143}Nd/^{144}Nd$ in extremely depleted mantle rocks. *Earth Planet. Sci. Lett.,* 135, 101–114.

Shengrong, L., Tsenmin, G., Nansheng, Ch. (1994) Platinum group elements - a possible diagnostic and metallogeny indicator. *7-th International Platinum symposium. Abstracts.* Moscow: Publishing House Moscow Contact, 61-62 (in Rus.).

Sun, S.-S., McDonough, W.F. (1989). Chemical and isotopic systematics of oceanic basalt: implications for mantle composition and processes. *Magmatism in ocean basins (Eds. A.D. Saunders., M.J. Norry). Geological Society Special Publication*, 42, 313–345.

Sun, S. S., Nesbitt, R.W. (1978) Petrogenesis of Archaean ultrabasic and basic volcanics: evidence from rare earth elements. *Contrib. Mineral. Petrol.*, 65 (5.), 301–325.

Van der Wal, D., Bodinier, J.-L. (1996) Origin of the recrystallisation front in the Ronda peridotite by km-scale pervasive porous melt flow. *Contrib. Mineral. Petrol.*, 122, 387–405.

Volokhov, I.M., Ivanov, V.M., Arnautov, N.V. et al. (1970) Mazhalyksky gabbro-pyroxenites-peridotite pluton (East Tannu-Ola, Tuva)./ *Problems of a petrology of the ultramafic and the mafic basic rocks*. Novosibirsk: Publishing House Nauka, Siberian branch, 130–145 (in Rus.).

Zhou, M.F., Robinson, P.T., Malpas, J. et al. (2005) REE and PGE geochemical constraints on the formation of dunites in the Luobusa ophiolite, Southern Tibet. *J. Petrol.* 46 (3), 615–639.

In: Platinum: Compounds, Production and Applications
Editors: L.Varennikov and E. Yedemsky
ISBN: 978-1-62257-939-6

Chapter 3

THE SYNTHESIS OF PLATINUM COMPLEXES OF CYCLIC PHOSPHINES AND BISPHOSPHINES

György Keglevich,[1] Péter Bagi,[1] Erika Bálint[1] and Tamás Körtvélyesi[2]

[1]Department of Organic Chemistry and Technology, Budapest University of Technology and Economics, Budapest, Hungary
[2]Department of Physical Chemistry and Material Science, University of Szeged, Szeged, Hungary

ABSTRACT

Novel platinum complexes incorporating phosphine ligands have been developed by us. On the one hand, platinum complexes of PtL_2Cl_2 type were synthesized, where L represents a 3-phospholene, phospholane, 1,2,3,6-tetrahydrophosphinine or 1,4-dihydrophosphinine moiety in differently substituted forms. In a few cases, the P-ligands were used in an optically active form. In the second group, 3-diphenylphosphino-substituted P-heterocycles, such as 3-Ph_2P-1-phenylphospholane, 3-Ph_2P-1-phenyl-1,2,3,6-tetrahydrophosphinine and 3-Ph_2P-1-phenyl-1,2,3,4,5,6-hexahydrophosphinine bisphosphines were synthesized to make available the corresponding *cis* chelate (ring) platinum complexes. Finally, bis(diphenylphospinomethyl)amines were utilized in the synthesis of other kind of ring platinum complexes. In the above cases, the phosphines and diphosphines were obtained from the corresponding phosphine oxide and bis(phosphine oxide) precursors after (double) deoxygenation by silanes. The cyclic bis(phosphine oxides) were obtained by the phospha-Michael reaction, while the bis(phosphinoxidomethyl)amines by the double Kabachnik–Fields (phospha-Mannich) reaction. The new platinum complexes are potential catalysts in homogenous catalytic processes. Five platinum complexes were calculated by DFT and semiempirical (PM6-DH2) methods.

Introduction

Homogeneous catalytic reactions are important tools in synthetic organic chemistry. Phosphines form an important group within organophosphorous compounds, as P(III)-transition metal complexes are widely applied catalysts in several homogeneous catalytic reactions, such as hydrogenation and hydroformylation [1-3]. In the last decades, several acyclic mono- and bidentate phospines, both achiral and chiral ones were developed, and after converting them to transition metal complexes they were successfully applied as catalysts [4]. Among the P-ligands, the P-heterocycles form a representative class [5]. *Pringle et al.* reported several 5-, 6- and 7-membered P-heterocycles [6], as well as 9-phosphabicyclononanes (Phobanes) [7] and 6-phospha-2,4,8-trioxaadamantanes [8]. Several bidentate heterocyclic P-ligands, such as DuPhos [9], PennPhos [10], BIPNOR [11] have also been described in the literature. The conversion of sterically congested 1-aryl-1*H*-phospholes with certain extent of aromaticity to transition metal complexes have been thoroughly investigated [12,13]. Recently, in our research group several achiral and optically active aryl-, amino- and alkoxy-substituted dibenzo[c.e][1,2]oxaphosphorines were synthetized and converted to the corresponding platinum complexes [14-22].

The purpose of this review is to summarize the most recent developments in the field of platinum complexes incorporating mono- and bidentate P-heterocycles, such as phospholenes, phospholanes, 1,2-dihydrophosphinines, 1,2,3,6-tetrahydrophosphinines and 1,2,3,4,5,6-hexahydrophosphinines, as well as bis(phosphinomethyl)amines.

1. Five- and Six-Membered P-Heterocycles as P-Ligands in Platinum Complexes

The racemic phenyl-, *n*-propyl and *n*-butyl-phospholene oxides (**1**) were converted to the corresponding phospholenes (**2**) by deoxygenation with trichlorosilane. The phosphines (**2**) so obtained were reacted with half equivalent of dichlorodibenzonitrileplatinum to afford the corresponding platinum complexes (**4**). Relative position of the phospholene moieties was evaluated by ^{31}P NMR spectroscopy. The $^{1}J_{Pt\text{-}P}$ coupling constants indicated that the *cis* complexes (**4**) were formed exclusively and the *trans* complexes were not present, not even in traces. The phosphine-borane complexes (**3**) were also synthesized by reacting the corresponding phosphines (**2**) with borane-dimethysulfide (Scheme 1) [23,24]. Phosphine-borane complexes can be regarded as phosphine precursors as the decomplexation can be easily achieved by secondary amines [25].

The enantiomers of 1-phenyl-, *n*-propyl and *n*-butyl-3-methyl-3-phospholene oxides (**1a-c**) can be obtained from racemic phospholene oxides (**1a-c**) by resolution with TADDOL (**5a**) or spiro-TADDOL (**5b**). During resolution, the resolving agent (**5**) and one of the enantiomers of phosholene oxide form a molecular complex (**1a-c • 5**) that precipitates and can be purified by recrystallization. The enantiomers (**1a-c**) were recovered from the complexes by column chromatography (Scheme 2) [24,26]. The resolution of aryl- and alkyl-3-methyl-3-phospholene oxides can also be performed via coordination complex formation with Ca^{2+} salts of dibenzoyl- and di-*p*-toluyltartaric acid [27].

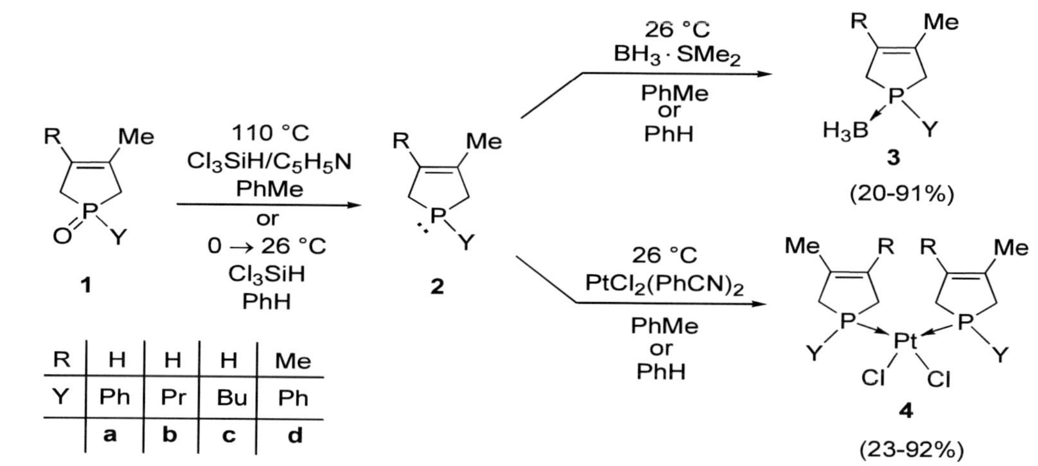

Scheme 1.

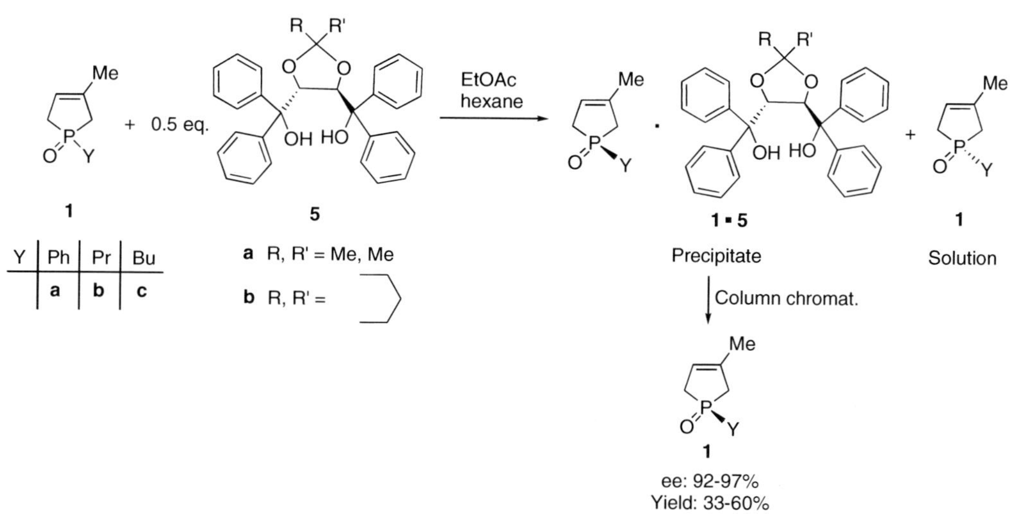

Scheme 2.

Optically active *cis*-1-phenyl-, 1-*n*-propyl and 1-*n*-butyl-3-methyl-3-phospholene-platinum complexes ((*R*,*R*)-**4a**, (*S*,*S*)-**4b**, (*S*,*S*)-**4c**) were also synthesized by an analogous reaction sequence shown in Figure 1 from optically active phospholene oxides ((*R*)-**1a**, (*S*)-**1b**, (*S*)-**1c**) (ee: 92-97) [23,24]. Retention of the P-asymmetric center was assumed [28].

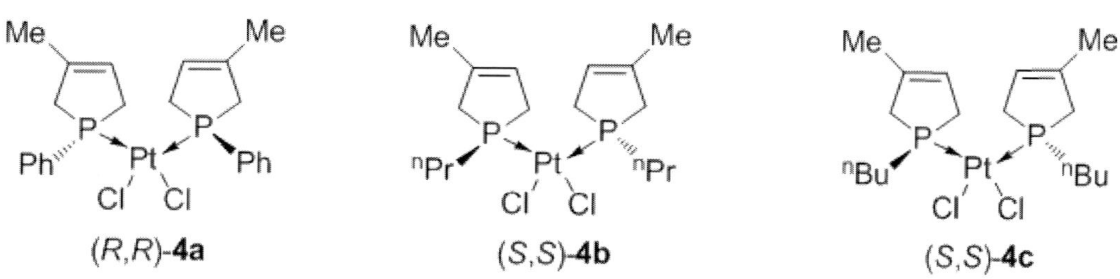

Figure 1.

The structures of the optically active (*R*,*R*) **4a** and (*S*,*S*) **4b** complexes were calculated by quantum chemical methods. The different groups attached to the P-atom determine the interactions within the molecules. In the (*R*,*R*)-1-phenyl-3-methyl-3-phospholene-platinum complex ((*R*,*R*)-**4a**) a π-CH_2 interaction between the suitable phenyl and 5-membered hetero rings stabilizes the molecule which can be considered a weak polar interaction (Figure 2).

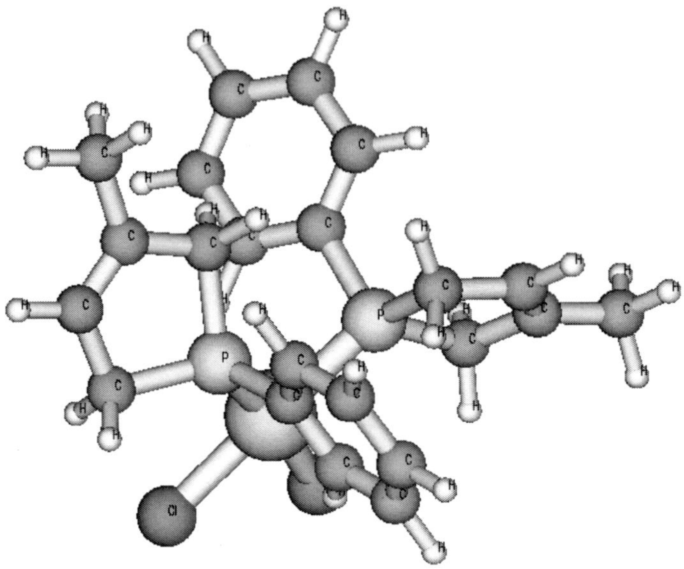

Figure 2. Stereostructure and geometries for *cis* bis(1-phenyl-3-methyl-3-phospholeno)-dichloroplatinum(II) (*R*,*R*)-**4a** calculated by the B3LYP/6-31G(d,p) (for atoms C, H, N, O, P and Cl) and LANL2DZ (for the Pt atom) methods. The geometries obtained by the B3LYP/6-31G(d,p) and LANL2DZ calculations are shown in "( )", while the geometries obtained by the PM6-DH2 method are in "{ }". Selected bond lengths (Å) and angles (°) are as follows: Pt-Cl 2.398, 2.397 (2.416, 2.415) {2.359, 2.360}, Pt-P 2.285, 2.285 (2.293, 2.293) {2.276, 2.276}, P-C(Ph) 1.820, 1.820 (1.834, 1.835) {1.848, 1.849}, P-C2 1.843, 1.852 (1.865, 1.862) {1.894, 1.885}, C2-C3 1.506, 1.509 (1.517, 1.514) {1.498, 1.503}, C3-C4 1.329, 1.328 (1.340, 1.341) {1.348, 1.347}, C4-C5 1.502, 1.499 (1.505, 1.509) {1.490, 1.494}, C5-P 1.846, 1.853 (1.865, 1.873) {1.895, 1.888}, Cl-Pt-P 85.44, 85.32 (85.57, 85.79) {92.42, 92.13}, P-C2-C3 105.40, 105.40 (105.61, 105.65) {102.10, 101.74}, C2-C3-C4 116.35, 115.89 (115.89, 116.36) {119.00, 119.00}, C3-C4-C5 118.14, 118.70 (118.86, 118.32) {119.90, 119.94}, C4-C5-P 104.40, 104.52 (104.59, 104.73) {101.51, 101.90}, Cl-Pt-P-C(Ph) 84.23, 83.78 (84.72, 85.80) {104.37, 104.16}, Pt-P-C2-C3 –136.08, 133.47 (–134.61, 132.18) {–139.34, 137.52}, P-C2-C3-C4 8.64, –7.55 (7.52, –6.48) {6.41, –6.91}, P-C2-C3-CH_3 –172.92, 174.06 (–173.96, 175.11) {–173.88, 173.82}.

Racemic 1-phenyl and 1-*n*-propyl-3-methyl-3-phospholene 1-oxides (**1a** and **1b**), as well as 1-phenyl-3,4-dimethyl-3-phospholene 1-oxide (**1d**) were converted to the corresponding phospholanes (**6**) by catalytic hydrogenation. The phospholane oxides (**6**) so obtained were subjected to deoxygenation and transformed to the corresponding borane (**8**) and platinum complexes (**9**) (Scheme 3) [23,24]. The stereostructure of the (*R*,*S*-*R*,*R*)-**9b** complex was also calculated by quantum chemical methods (Figure 3). No significant intramolecular interactions could be found due to the unsuitability of the propyl group.

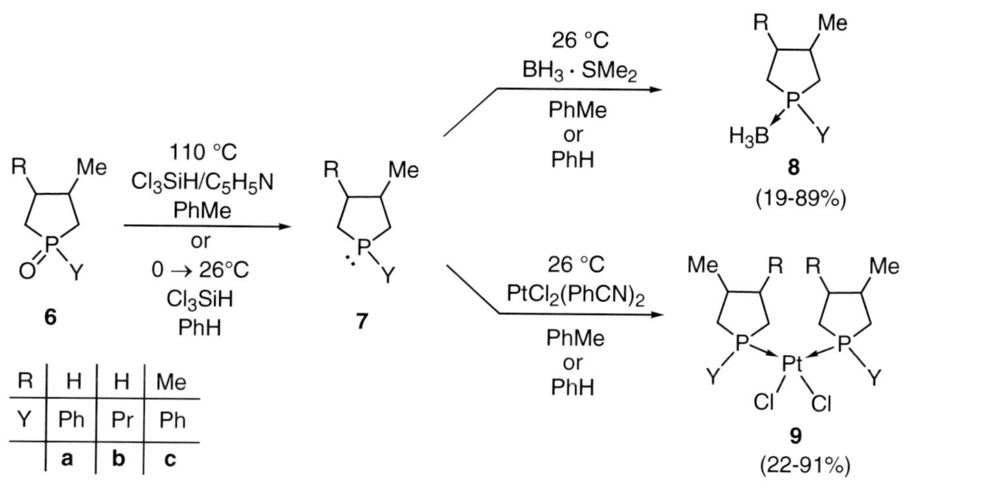

Scheme 3.

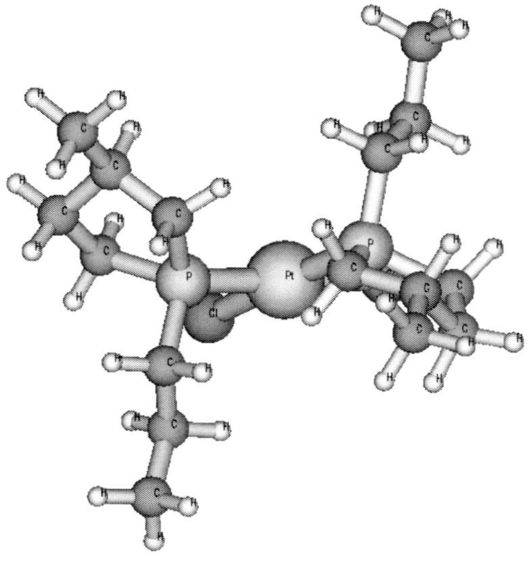

Figure 3. Perspective view and geometries for *cis* bis(1-propyl-3-methylphospholano)-dichloroplatinum(II) (*R,S-R,R*)-9b calculated by the BHANDHLYP/6-31G(d,p)–LANL2DZ methods. The geometries obtained by the B3LYP/6-31G(d,p)–LANL2DZ methods are in "( ), while the geometries obtained by the PM6-DH2 method are in "{ }". Selected bond lengths (Å) and angles (°) are as follows: Pt-Cl 2.399, 2.400 (2.419, 2.419) {2.364, 2.370}, Pt-P 2.290, 2.292 (2.297, 2.298) {2.274, 2.270}, P-C2 1.856, 1.851 (1.874, 1.869) {1.889, 1.890}, C2-C3 1.527, 1.535 (1.547, 1.545) {1.540, 1.538}, C3-C4 1.529, 1.529 (1.540, 1.541) {1.552, 1.552}, C4-C5 1.538, 1.530 (1.537, 1.539) {1.525, 1.523}, P-C5 1.842, 1.848 (1.861, 1.867) {1.882, 1.883}, P-C1' 1.837, 1.837 (1.856, 1.856) {1.886, 1.887}, C1'-C2' 1.527, 1.527 (1.536, 1.536) {1.532, 1.533}, C2'-C3' 1.522, 1.522 (1.533, 1.533) {1.528, 1.528}, Cl-Pt-Cl 90.2 (89.6) {81.50}, Cl-Pt-P 85.3, 85.7 (85.4, 85.9) {90.5, 92.8}, Pt-P-C2 120.1, 119.6 (120.2, 119.6) {122.1, 118.9}, Pt-P-C5 116.8, 116.8 (117.0, 117.1) {114.9, 118.7}, Pt-P-C1' (113.9, 113.8) {105.6, 105.4}, P-C2-C3 106.4, 106.1 (106.6, 106.3) {103.1, 103.9}, C2-C3-C4 105.8, 106.4 (105.8, 106.5) {109.2, 111.0}, C3-C4-C5 108.3, 107.4 (107.6, 108.4) {110.9, 112.0}, Cl-Pt-P-C2 –147.5, –152.5 (–147.2 –151.7) {–133.4, –141.3}, Pt-P-C1'-C2' –49.1, –58.5 (–47.9, –59.3) {–36.7, -26.5}, Pt-P-C2-C3 140.4, 114.3 (140.2, 114.4) {145.1, 120.2}, P-C2-C3-C4 –38.5, 34.2 (–38.2, 34.1) {–33.7, 25.8}, C2-C3-C4-C5 48.3, –48.7 (48.3, –48.6) {41.3, –41.3}, P-C1'-C2'-C3' 179.1, –179.2 (178.9, –179.5) {172.0, 180.0}.

A phospholane oxide having an exocyclic $P(O)Ph_2$ function in position 3 (**11**) was synthetized by the Michael addition of diphenylphosphine oxide to the double-bond of 1-phenyl-2-phospholene 1-oxide (**10**) in the presence of NaOMe/MeOH (Scheme 4) [29].

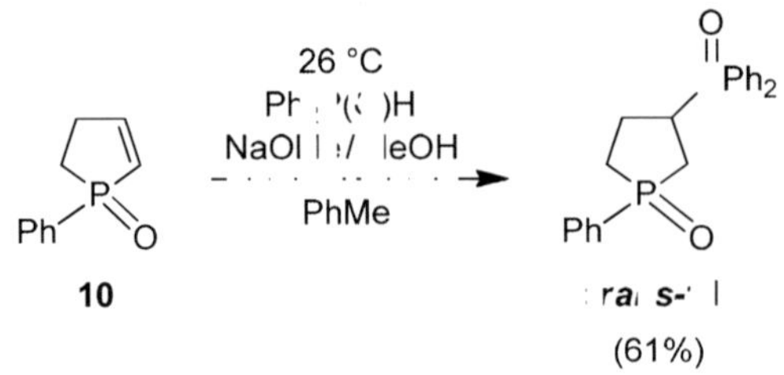

Scheme 4.

The double deoxygenation of *trans*-3-diphenylphosphinoxido-1-phenylphospholane 1-oxide (**11**) was achieved by trichlorosilane. The diphosphine (**12**) so obtained was converted to the corresponding borane (**13**) and *cis* chelate-platinum complex (**14**) by reaction with two equivalents of borane-dimethylsulfide, or one equivalent of dichlorodibenzonitrileplatinum (Scheme 5) [29]. Platinum(II) complex **14** was calculated by the B3LYP/6-31G(d)-LANL2DZ, the BHANDHLYP/6-31G(d,p), as well as the PM6-DH2 methods. According to the BHANDHLYP/6-31G(d,p) calculation, the distances between $C_{\alpha'}$ and $C_{\alpha''}$ and between $C_{\beta'}$ and $C_{\beta''}$ were found to be 2.93 and 3.57 Å respectively. The same values obtained by B3LYP/6-31G(d)-LANL2DZ calculations were 2.95 and 3.55 Å, respecively, while PM6-DH2 led to 3.02 and 3.52 Å, respectively. As it can be seen in Figure 4, the Ph groups in the PPh_2 moiety have a deformed π-π stacking (T-shape) interaction which is better shown by BHANDHLYP/6-31G(d,p) and LANL2DZ calculations than by the other two methods. A significiant difference was found in the phenyl conformations by the three methods. The C_β(Ph)-C_α(Ph)-P-Pt torsion angles were found to be –37.89°/–8.33°, 29.85°/–4.11° and 12.16°/–4.39°, respectively, according to the three methods mentioned above. The Pt-P and Pt-P1 distances were found to be 2.279/2.288 Å, 2.303/2.282 Å and 2.290/2.278 Å, respectively, by the three methods. The BHANDHLYP functional handles the H-bonds and dispersion interactions in geometry much better than the B3LYP method.

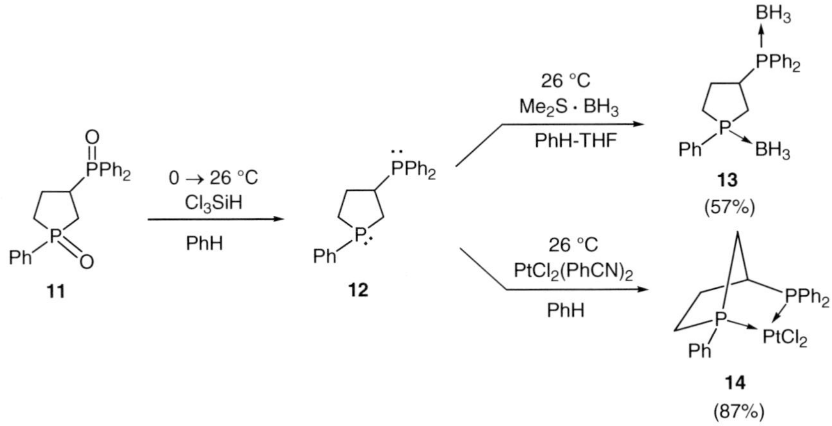

Scheme 5.

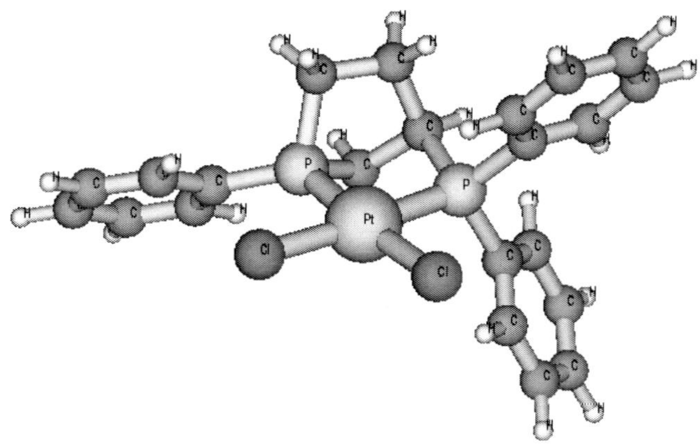

Figure 4. Perspective view and selected geometries for ring platinum(II) complex **14** obtained by the BHANDHLYP/6-31G(d)–LANL2DZ methods. The geometries obtained by the B3LYP/6-31G(d)–LANL2DZ calculations are shown in "( )", while the geometries obtained by the PM6-DH2 method are in "{ }". Selected bond lengths (Å) and angles (°) are as follows: Pt-Cl 2.385, 2.397 (2.418, 2.401) {2.358, 2.354}, Pt-P 2.279 (2.303) {2.290}, Pt-P1 2.288 (2.282) {2.278}, P-C(Ph) 1.815, 1.821 (1.837, 1.836) {1.845, 1.842}, P1-C(Ph), 1.803 (1.817) {1.831}, P1-C2 1.839 (1.857) {1.907}, C2-C3 1.536 (1.544) {1.528}, C3-C4 1.552 (1.561) {1.546}, C4-C5 1.552 (1.561) {1.536}, C5-P1 1.841 (1.862) {1.903}, Cl-Pt-Cl 91.70 (90.72) {81.85}, Cl-Pt-P1 92.94 (91.99) {95.88}, Cl-Pt-P 91.05 (92.00) {95.78}, Pt-P1-C(Ph) 126.41 (104.53) {127.45}, Pt-P-C(Ph) 114.73, 118.17 (114.04, 120.08) {110.60, 117.67}, P1-C2-C3 98.35 (98.34) {97.34}, C2-C3-C4 106.36 (106.02) {108.52}, C3-C4-C5 110.26 (110.25) {112.19}, C4-C5-P1 103.87 (104.15) {103.07}, Cl-Pt-P1-C(Ph) –25.29 (–22.29) {–14.66}, Cl-Pt-P1-C2 –156.71 (–152.40) {–146.14}, Pt-P1-C2-C3 –58.06 (–61.13) {–65.78}, P1-C2-C3-C4 –51.37 (–52.77) {–50.28}, C2-C3-C4-C5 29.03 (33.06) {40.04}, C3-C4-C5-P1 7.99 (3.34) {–6.64}, Cβ(Ph)-Cα(Ph)-P1-Pt 59.93 (61.75) {40.16}, Cβ(Ph)-Cα(Ph)-P-Pt –37.89, (–8.33 29.85, –4.11) {12.16, –4.39}.

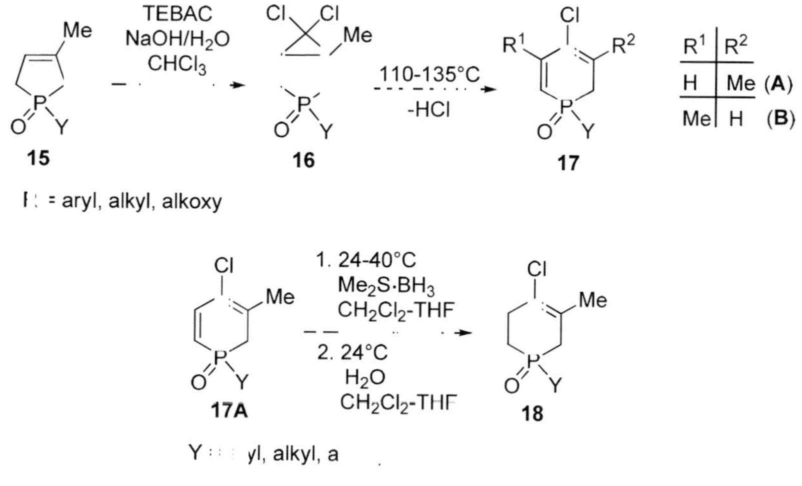

Scheme 6.

The 5-membered P-heterocycles are versatile starting materials for several P-heterocycles with larger ring system. 3-Phospholene oxides (**15**) can be converted to 6-membered P-heterocycles (**16-17**) by ring enlargement. The first step of this procedure is the addition of dichlorocarbene on the double bond of the 3-phospholene oxide (**15**) that leads to the corresponding 3-phosphabicyclo[3.1.0]hexane 3-oxides (**16**). The thermolytic opening of the cyclopropane ring affords a 3:1 mixture of the two double-bond isomers of the 1,2-dihydrophosphinine oxides (**17A** and **17B**) [30]. Selective reduction of the α,β-double-bond of the 3-methyl-1,2-dihydrophosphinine oxides (**17A**) by borane-dimethylsulfide provides the corresponding 1,2,3,6-tetrahydro derivatives (**18**) (Scheme 6) [31].

The dichlorocyclopropanation of 3,4-dimethyl-3-phospholene oxides (*e.g.* **1d**) was followed by an immediate ring opening to give 1,2-dihydrophosphinine oxides (*e.g.* **20**) which underwent insertion of a second dichlorocarbane unit and then a ring opening to result in 4-dichloromethylene-1,4-dihydrophosphinine oxides (*e.g.* **22**) [32].

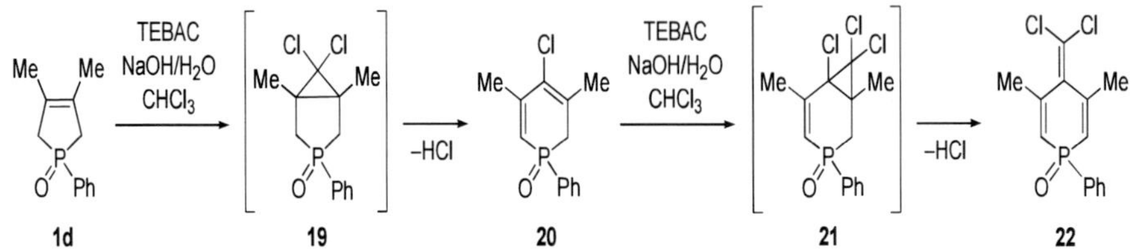

Scheme 7.

The 5-methyl-4-chloro-1,2,3,6-tetrahydrophosphine oxide (**18**) and the 4-dichloromethylene-3,5-dimethyl-1-phenyl-1,4-dihydrophosphinine oxide (**22**) were subjected to deoxygenation by trichlorosilane in the presence of pyridine. The phosphines (**23** and **24**) so obtained were converted immediately to the corresponding platinum complexes (**26** and **27**) by reaction with dichlorodibenzonitrileplatinum. The 5-methyl-4-chloro-1,2,3,6-tetrahydrophosphine-borane complex (**25**) was also prepared (Scheme 8) [23].

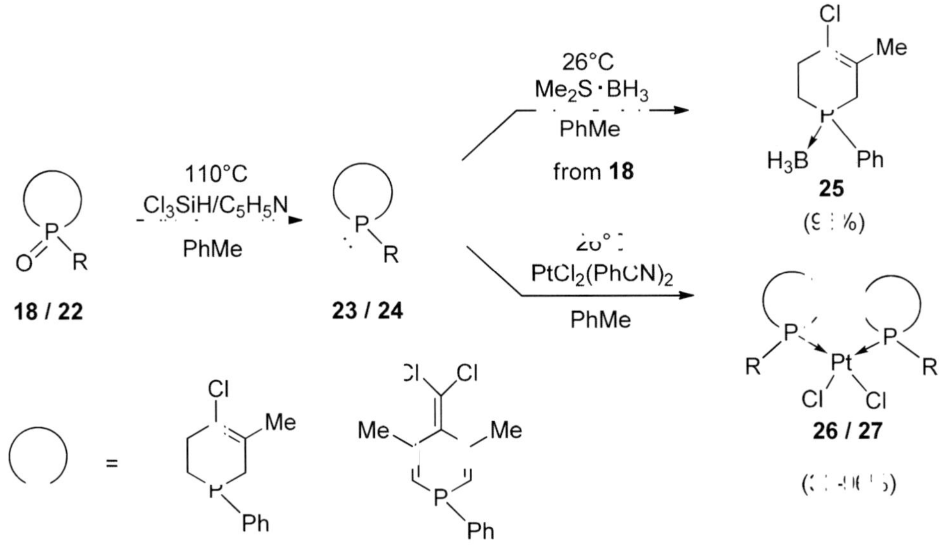

Scheme 8.

1,2-Dihydrophosphinine oxides (**17A**) can be converted to 1,2,3,6-tetrahidrophosphinine oxides with exocyclic P=O function in position 3 (**28**) by the Michael addition of >P(O)H species to the electron-poor α,β-double-bond (Scheme 9). The reaction took place in a diastereoselective manner resulting in 3-phosphono- and 3-phosphinoxido-1,2,3,6-tetrahidrophosphinine oxides (**28**) having the >P(O) moiety and the oxygen atom of the ring P=O in position *cis* [33-35].

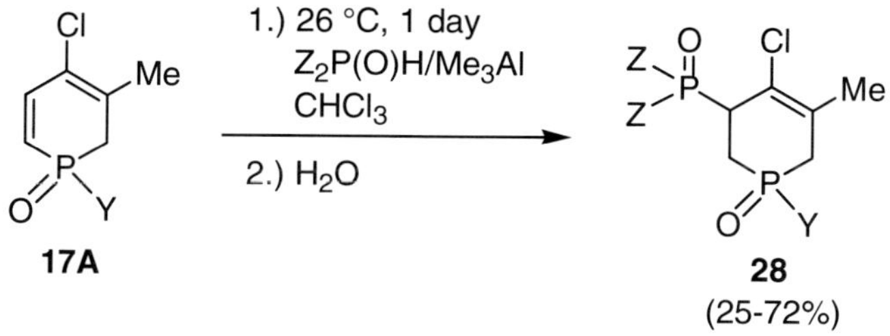

Y	Ph	Ph	Ph	4-MeC_6H_4	4-MeC_6H_4	Et	EtO	EtO	EtO	EtO
Z	Ph	MeO	EtO	Ph	EtO	EtO	Ph	MeO	EtO	$PhCH_2O$
	a	**b**	**c**	**d**	**e**	**f**	**g**	**h**	**i**	**j**

Scheme 9.

The 3-diphenylphosphinoxido-1-phenyl-1,2,3,6-tetrahidrophosphinine oxide (**28a**) was subjected to double deoxygenation and the diphosphine (**29**) so obtained was converted to the corresponding borane (**30**) and *cis* chelate platinum complex (**31**) (Scheme 10) [36].

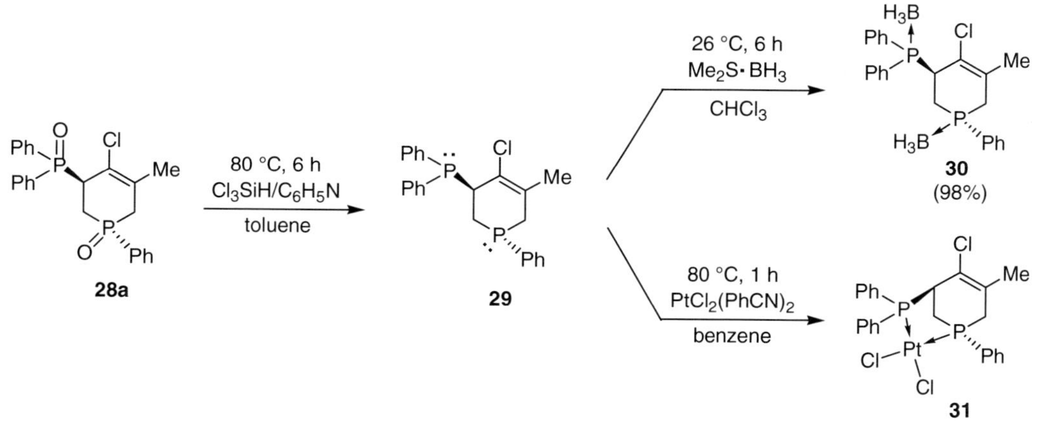

Scheme 10.

The BHANDHLYP and B3LYP calculations predicted a very similar structure shown in Figure 5. The PM6-DH2 calculations taking into account dispersions and H-bonds gave similar bond length as the density functional calculations, only the geometry of the phenyl rings were different. The effect of the C(Ar)-H...Cl interactions should be noted.

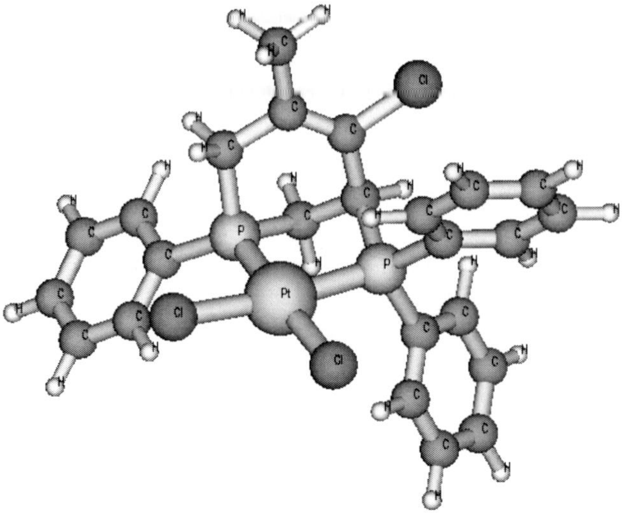

Figure 5. Perspective view and geometries for ring platinum complex **31** obtained by the BHANDHLY/6-31G(d,p)–LANL2DZ methods. The results of B3LYP/6-31G(d,p)–LANL2DZ calculations and those of the PM6-DH2 methods are shown in "( )" and "{ }", respectively. Selected bond lengths (Å) and angles (°) are as follows: Pt-Cl 2.391, 2.395 (2.391, 2.395) {2.355, 2.358}, Pt-P1 2.283 (2.290) {2.283}, Pt-P2 2.267 (2.273) {2.270}, P1-C(Ar) 1.810, 1.820 (1.835, 1.826) {1.839, 1.838}, P2-C(Ar) 1.809 (1.824) {1.834}, P1-C 1.886 (1.911) {1.943}, C-C(Cl) 1.511 (1.516) {1.490}, C(Cl)-C(Me) 1.335 (1.347) {1.350}, C(Me)-C 1.515 (1.524) {1.494}, C-P2 1.825 (1.844) {1.896}, P2-CH2 1.832 (1.850) {1.899}, P1-Pt-P2 86.41 (86.77) {85.52}, Cl-Pt-P2 90.52 (90.95) {97.38}, Pt-P1-C(Ar) 114.58, 116.90 (116.81, 119.89) {110.86, 114.81}, Pt-P2-C(Ar) 123.67 (124.03) {124.61}, Cl-Pt-Cl 92.34 (91.67) {81.47}, P2-CH2-CH 105.38 (105.36) {104.56}, P1-CH-C(Cl) 112.17 (112.34) {108.07}, CH-C(Cl)-C(Me) 127.44 (127.53) {127.91}, C(Cl)-C(Me)-CH2 123.61 (123.40) {125.09}, C(Me)-CH2-P2 113.93 (114.22) {113.44}, Pt-P1-CH-C(Cl) 75.62 (76.28) {90.21}, P1-CH-C(Cl)-C(Me) –90.00 (–89.44) {–77.44}, CH-C(Cl)-C(Me)-CH2 –0.03 (0.01) {0.66}, C(Cl)-C(Me)-CH2-P2 10.78 (10.36) {–6.36}, P1-CH-CH2-P2 60.36 (60.07) {53.97}, Pt-P1-C(Ar)-C2(Ar) 146.04, –34.59 (–2.24, –21.84) {–3.71, 3.12}, Pt-P2-C(Ar)-C2(Ar) –179.15 (–179.32) {168.60}.

The 3-phosphono- and 3-phosphinoxido-1,2,3,6-tetrahydrophosphinine oxides (**28a-c,f,g**) were converted to the corresponding 1,2,3,4,5,6-hexahydrophosphinine oxides (**32**) by catalytic hydrogenation, that involved saturation of the double-bond and hydrogenolysis of the C–Cl bond (Scheme 11) [37,38].

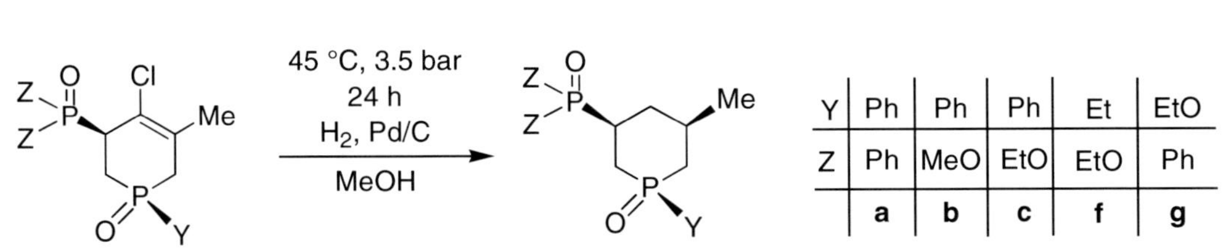

28 **32** (45-92%)

Scheme 11.

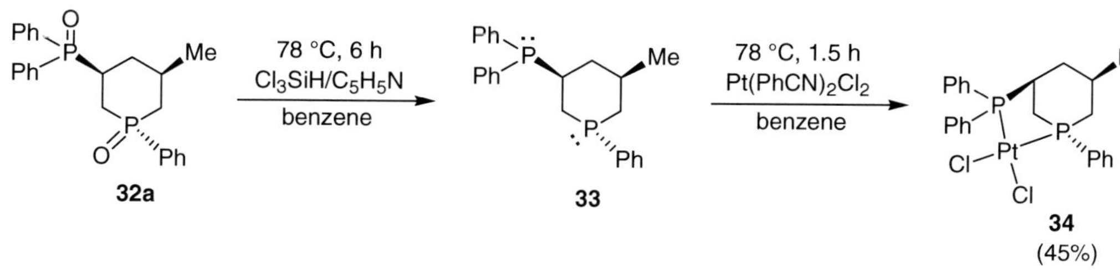

Scheme 12.

The 3-diphenylphosphinoxido-1-phenyl-1,2,3,4,5,6-hexahydrophosphi-nine oxide (**32a**) was deoxygenated with trichlorosilane. The reaction of the resulting diphosphine (**33**) with dichlorodibenzonitrileplatinum led to *cis* chelate complex (**34**) (Scheme 12) [39].

2. The Double Kabachnik–Fields Reaction; Development of Novel Bidentate P-Ligands

α-Aminophosphonates and related derivatives are in the focus these days due their versatile bioactivity [40-42]. A convenient approach to α-aminophosphonates is the condensation of an amine, an oxo compound (aldehyde or ketone) and a >P(O)H species, such as dialkyl phoshites or secondary phosphine oxides that is called the Kabachnik–Fields (or phospha-Mannich) reaction [43,44]. This three-component condensation can be accomplished in many ways [45]. A new trend has been to find appropriate catalysts allowing efficient syntheses [46]. However, it was found that especially under solventless conditions, there is no need for any catalyst [47,48]. The efficiency could be further increased by the microwave (MW) technique [49-51]. A novel preparation of α-aminophosphonates involves the substitution of α-hydroxyphosphonates by amines. This reaction takes place with an unexpected ease as a consequence of a neighbouring group effect [52].

Bis(phosphonomethyl)amines may be interesting building blocks in synthetic organic chemistry. A few liphophilic derivatives were prepared by the double Kabachnik–Fields condensation [53]. Bis(phosphinoxido) derivatives were synthesized in another way by an unexpected methylene insertion reaction into a P–N bond followed by oxidation of the P(III) center [54].

Extending the scope of the double Kabachnik-Fields (or bis(phospha-Mannich)) reaction, the condensation of an *n*-propyl- and *n*-butylamine, cyclohexylamine, benzylamine and 4-methoxyaniline, two equivalents of paraformaldehyde and two equivalents of a >P(O)H species, such as dialkyl phosphites, ethyl phenyl-*H*-phosphinate or secondary phosphine oxide were investigated. The reactions were carried out under MW conditions at 100 °C for 1 h in the absence of any solvents and catalysts. The bis(>P(O))-products (**35a-b**, **36a-b**, **37a-e**, **38a-e** and **39a-d**) were obtained in 76-98% yields after flash column chromatography (Scheme 13) [55-57].

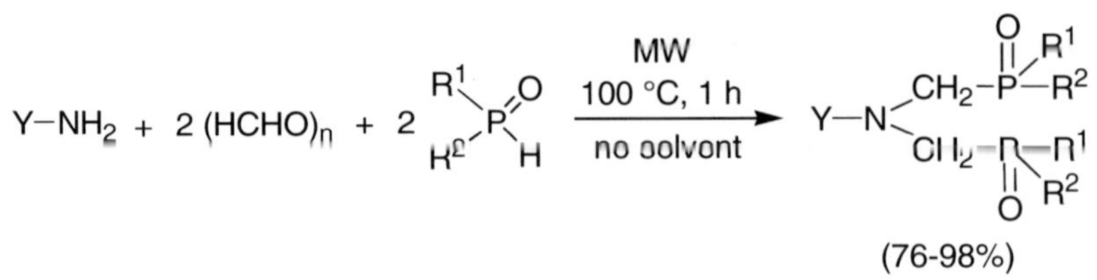

Y = ^{n}Pr (**35**), ^{n}Bu (**36**), ^{c}Hex (**37**), Bn (**38**), 4-MeOPh (**39**)

R^1	OMe	OEt	OBu	OBn	OEt
R^2	OMe	OEt	OBu	OBn	Ph
	a	**b**	**c**	**d**	**e**

Scheme 13.

Using diphenylphosphine oxide as the P-component and *n*-propyl-, *n*-butylamine, cyclohexylamine, benzylamine and arylamines along with paraformaldehyde in an apropriate molar ratio at 100 °C for 1-1.5 h, the corresponding bis(phosphine oxides) **(40a-g)** were obtained in 55-95% yields after flash column chromatography. Due to the heterogeneity of the reaction mixture, it was better to carry out the condensation in acetonitrile (Scheme 14) [56-57].

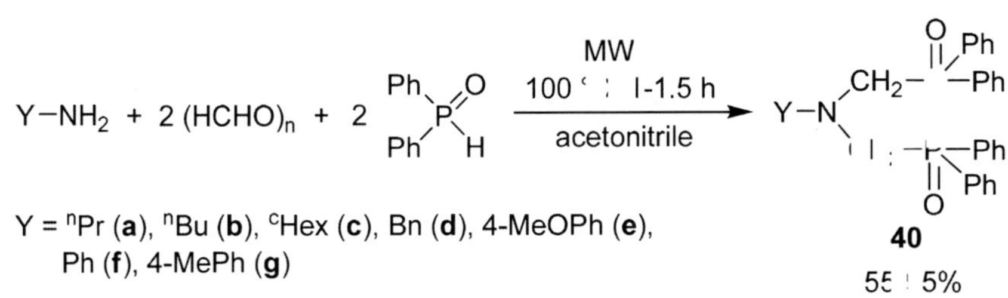

Scheme 14.

The bis(diphenylphosphinoxidomethyl)amines **(40a-g)** were then subjected to a double deoxygenation applying phenylsilane [28] in benzene at 80 °C. The diphosphines **(41a-g)** so obtained were reacted with a half equivalent of dichlorodibenzonitrileplatinum to provide the corresponding bis(phosphine) platinum(II) complexes **(42a-g)** in a pure form after filtration (Scheme 15).

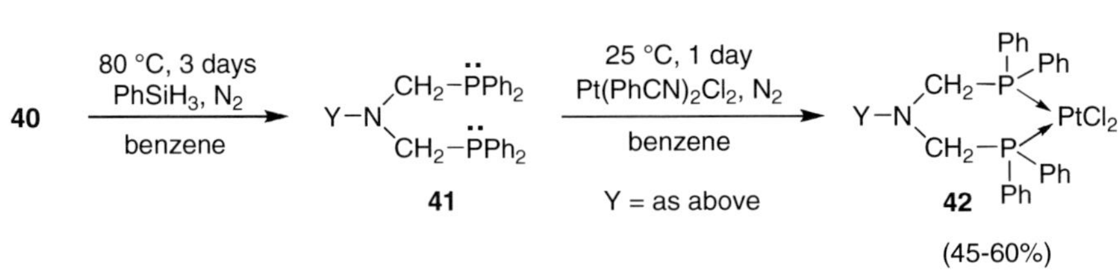

Scheme 15.

The results on bis(diphenylphosphinomethyl)-cyclohexylamine-platinum complex **42c** obtained by the obtained by TPSS/6-31G(d,p) and LANL2DZ calculations, by the B3LYP/6-31G(d,p) and LANL2DZ methods, or by the PM6-DH2 calculation are shown in Figure 6. The six-membered ring including a Pt atom has a deformed chair conformation. The distances between the C(α)...C(β') and C(α')...C(β) are 3.35 Å, 3.23 Å and 3.55 Å, 3.44 Å respectively, according to TPSS /6-31G(d,p)-LANL2DZ and PM6-DH2 calculations. The same values were 3.22 Å and 3.23 Å as a result of B3LYP/6-31G(d,p)-LANL2DZ calculations. The C-N-C angles are 114.2°/115.6°/ 111.5° and 112.2°/114.7°/111.8° according to the previous two methods. This parameter was 114.1°/115.9°/112.0° according to the B3LYP/6-31G(d,p)-LANL2DZ optimization. The CH_2-N-C1-C2 torsion angles are 158.1°/–70.5° and 158.8°/–72.2° according to the TPSS/6-31G(d,p)-LANL2DZ and PM6-DH2 calculations. The B3LYP/6-31G(d,p)-LANL2DZ optimization provided a torsion angle of 156.8°/–70.4°. A significant difference was found in the Pt-P distance obtained by these three methods: XRD resulted 2.29 Å, while B3LYP/ 6-31G(d,p)-LANL2DZ and PM6-DH2 predicted the closest distance with 2.273/2.274 Å and 2.284/2.281 Å, respectively.

The Pt-P distances obtained by the TPSS calculation are larger than those obtained by the other two methods. The PM6-DH2 semiempirical method and the DFT calculations predict a very similar geometry except the conformation of the phenyl groups (see the data listed in Figures 2 and 6).

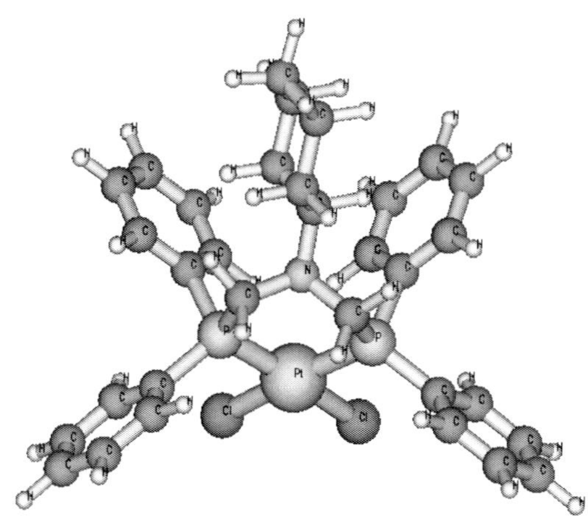

Figure 6. Perspective view and geometries for bis(diphenylphosphinomethyl)-cyclohexylamine-platinum complex **42c** obtained by the TPSS/6-31G(d,p)–LANL2DZ calculations. The results of B3LYP/6-31G(d,p)–LANL2DZ methods are shown in "( )", while the data obtained by the PM6-DH2 method can be seen in "{ }". Selected bond lengths (Å) and angles (°) are as follows: Pt-P 2.292, 2.294 (2.273, 2.274) {2.284, 2.281} P-CH_2 1.882, 1.873 (1.877, 1.868) {1.898, 1.901}, CH_2-N 1.462, 1.463 (1.455, 1.456) {1.489, 1.482}, P-$C_{1'}$ 1.834, 1.837, 1.840, 1.837 (1.836, 1.839, 1.836, 1.833) {1.849, 1.850,1.850, 1.848}, N-C_1 1.4 (1.491) {1.507}, C_1-C_2 1.547 (1.544) {1.542}, C_2-C_3 1.543 (1.538) {1.534}, C_3-C_4 1.540 (1.536) {1.535}, Pt-P-$C_{1'}$ 113.60, 114.14, 116.26, 112.93 (113.64, 113.93, 116.28, 113.05) {107.89, 110.54, 107.75, 112.32}, Pt-P-CH_2 115.25, 113.32 (115.10, 113.21) {119.01, 115.95}, P-CH_2-N 112.60, 112.07 (113.75, 113.46) {96.48, 93.91}, CH_2-N-C_1 115.64, 114.193 (115.93, 114.44) {114.17, 112.20}, CH_2-N-CH_2 111.50 (112.04) {111.81}, N-C_1-C_2 111.26 (111.83) {108.84}, C_1-C_2-C_3 110.84 (111.15) {109.89}, C_2-C_3-C_4 111.46 (111.66) {111.35}, Cl-Pt-P-$C_{1'}$ 52.29, –72.03, –46.46, 78.08 (51.17, –73.16, –46.44, 77.87) {54.46, –65.60, –46.66, 74.57}, Cl-Pt-P-CH_2 167.27, –162.0 (165.93, –161.51) {174.83, –167.21}, Pt-P-$C_{1'}$-$C_{2'}$ 54.43, 6.66, –27.40, –33.77 (54.10, 7.94, –26.95, –37.89) {8.88, –3.93, –6.39, –15.14}, Pt-P-CH_2-N 43.49, –50.99 (44.31, –51.09) {33.54, –48.35}, CH_2-N-C_1-C_2 –70.48, 158.12 (–70.41, 156.81) {–72.23, 158.76}, N-C_1-C_2-C_3 –175.07 (–175.10) {–175.00}, C_1-C_2-C_3-C_4 –55.99 (–55.48) {–56.50}.

The platinum complexes **42c-e** and **42g** were tested as catalysts in the hydroformylation of styrene. The platinum-containing *in situ* catalysts formed either from species **42c-e** or **42g** and tin(II) chloride were used under standard oxo-conditions. In general, the branched and the linear formyl regioisomers (2-phenylpropanal and 3-phenylpropanal, respectively) may be formed in the hydroformylation of styrene. The *in situ* formed platinum-diphosphine-tin(II)chloride systems proved to be active catalysts providing, interestingly, the branched aldehyde in high regioselectivities.

Bis(diphenylphosphinomethyl)-cyclohexylamine (**41c**) was converted to the bis(diphenylphosphinomethyl)-cyclohexylamine-borane complex (**43**) with borane dimethyl sulfide (Scheme 17). It is known that the phospine-boranes are precursors of the corresponding phosphines, as on treatment with secondary amines, the free phosphines can be liberated from them [25].

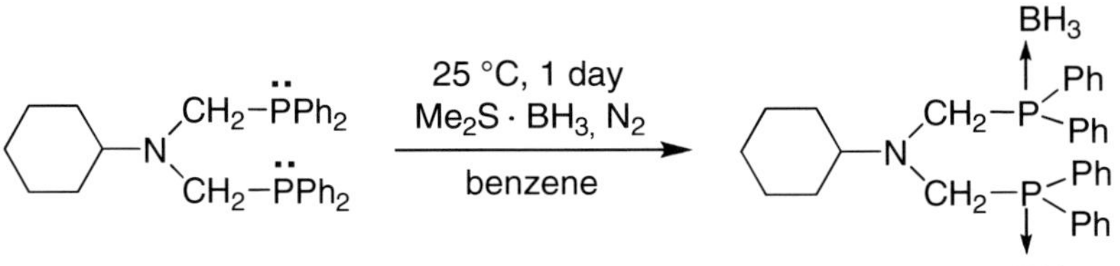

41c **43**

Scheme 17.

3. Experimental for the Theoretical Calculations

The structure of the Pt complexes with cyclic P-ligands were studied by DFT (Density Functional Theory) methods implemented in Gaussian '03 [58]. The initial geometries were built up by PCMODEL [59] and optimized by PM6 Hamiltonian [60] in MOPAC2007 and MOPAC2009 [61]. A few structures were optimized by PM6-DH2 which is a PM6 method with correction to dispersion and H-bonding [62]. B3LYP, BHANDHLYP and TPSSTPSS functionals were applied in the calculations for C, H, N, O, P, Cl atoms. The latter two methods have been developed for the description of dispersion, more precisely the H-bonds and the interactions in organometallic compounds. The LANL2DZ ECP basis set was applied for the Pt atom. In the minima no imaginary frequencies were found.

Acknowledgments

The above project was supported by the Hungarian Scientific and Research Fund (OTKA K83118. T. K. is grateful for the Hungarian Supercomputer Centre, NIIF and HPC Szeged for the computational possibilities. T. K. is indebted for the TAMOP-4.2.1/B-09/1/KONV-2010-0005.

References

[1] Mathey, F. Application of phosphorous heterocycles in homogeneous catalysis. In: Mathey, F. ed. *Phosphorous-Carbon Heterocyclic Chemistry: The Rise of a New Domain*. Ch. 7, Amsterdam, Pergamon, 2001, pp. 753-772.

[2] Brunner, H.; Zettlmeier, W. *Handbook of enantioselective catalysis with transition metal compounds*. Weinheim, VCH, 1993.

[3] Botteghi, C.; Marchetti, M.; Paganelli, S. In: Beller, M.; Bolm, C. eds. *Transition metals for organic synthesis*. Weinheim, Wiley-VCH, 1998, Vol. 2, p. 25.

[4] Agbossou, F.; Carpentier, J.-F.; Mortreux, A. Asymmetric Hydroformylation. *Chem. Rev.* 1995, 95, 2485-2506.

[5] Kollár, L.; Keglevich, G. P-Heterocycles as Ligands in Homogeneous Catalytic Reactions *Chem. Rev.* 2010, 110, 4257-4302.

[6] Baber, R. A.; Haddow, M. F.; Middleton, A. J.; Orpen, A. G.; Pringle, P. G.; Haynes, A.; Williams, G. L.; Papp, R. Ligand stereoelectronic effects in complexes of phospholanes, phosphinanes, and phosphepanes and their implications for hydroformylation catalysis. *Organometallics* 2006, 26, 713-725.

[7] Carreira, M.; Charernsuk, M.; Eberhard, M.; Fey, N.; Ginkel, R. v.; Hamilton, A.; Mul, W. P.; Orpen, A. G.; Phetmung, H.; Pringle, P. G. Anatomy of Phobanes. Diastereoselective synthesis of the three isomers of n-butylphobane and a comparison of their donor properties. *J. Am. Chem. Soc.* 2009, 131, 3078-3092.

[8] Downing, J. H.; Floure, J. l.; Heslop, K.; Haddow, M. F.; Hopewell, J.; Lusi, M.; Phetmung, H.; Orpen, A. G.; Pringle, P. G.; Pugh, R. I.; Zambrano-Williams, D. General routes to alkyl phosphatrioxaadamantane ligands. *Organometallics* 2008, 27, 3216-3224.

[9] Clark, T.; Landis, C. Recent developments in chiral phospholane chemistry. *Tetrahedron: Asymmetry* 2004, 15, 2123-2137.

[10] Jiang, Q.; Jiang, Y.; Xiao, D.; Cao, P.; Zhang, X. Highly enantioselective hydrogenation of simple ketones catalyzed by a Rh–PennPhos complex. *Angew. Chem. Int. Ed.* 1998, 37, 1100-1103.

[11] Robin, F.; Mercier, F.; Ricard, L.; Mathey, F.; Spagnol, M. BIPNOR: A new, efficient bisphosphine having two chiral, nonracemizable, bridgehead phosphorus centers for use in asymmetric catalysis. *Chem. – Eur. J.* 1997, 3, 1365-1369.

[12] Csók, Z.; Keglevich, G.; Petőcz, G.; Kollár, L. Platinum complexes of phospholes with reduced pyramidal character from steric crowding. *Inorg. Chem.* 1999, 38, 831-833.

[13] Csók, Z.; Keglevich, G.; Petőcz, G.; Kollár, L. Coordination chemistry and hydroformylation activity of platinum complexes containing 1-aryl-phospholes. *J. Organomet. Chem.* 1999, 586, 79-84.

[14] Keglevich, G.; Kerényi, A.; Körtvélyesi, T. Synthesis of novel heterocyclic P-ligands and their application in transition metal complexes. In: Varga, B.; Kis, L. eds. *Transition Metal Chemistry: New Research*. New York, Nova Science Publishers, 2008, pp. 83-99.

[15] Keglevich, G.; Kerényi, A. Synthesis of dibenzo[c.e] [1,2]oxaphosphiranes and their application in platinum(II) complexes. *Trends in Organic Chemistry* 2008, 12, 73-77.

[16] Keglevich, G.; Szelke, H.; Kerényi, A.; Imre, T.; Ludányi, K.; Dukai, J.; Nagy, F.; Aranyi, P. 2-aryl-dibenzo-1,2-oxaphosphorine as a ligand in borane and in Pt(II) complexes. *Heteroatom Chem.* 2004, 15, 459-463.

[17] Keglevich, G.; Szelke, H.; Kerényi, A.; Kudár, V.; Hanusz, M.; Simon, K.; Imre, T.; Ludányi, K. New chiral P-ligands: P-amino- and P-cycloalkoxy dibenzo[c.e][1,2]oxaphosphorines. *Tetrahedron: Asymmetry* 2005, 16, 4015-4021.

[18] Keglevich, G.; Kerényi, A.; Mayer, B.; Körtvélyesi, T.; Ludányi, K. Platinum(II) complexes of 2-alkoxy-dibenzo[c.e][1,2]oxaphosphorines. *Transit. Metal Chem.* 2008, 33, 505-510.

[19] Keglevich, G.; Kerényi, A.; Szelke, H.; Imre, T. Bis(dibenzo [c.e][1,2]oxaphosphorino)dichloroplatinum complexes. *Transit. Metal Chem.* 2006, 31, 306-309.

[20] Keglevich, G.; Kerényi, A.; Szelke, H.; Ludányi, K.; Körtvélyesi, T. 2-Diphenylphosphino-2'-hydroxybiphenyl-based beta-ligands and their platinum(II) complexes. *J. Organomet. Chem.* 2006, 691, 5038-5044.

[21] Kerényi, A.; Balassa, A.; Körtvélyesi, T.; Ludányi, K.; Keglevich, G. Synthesis and complexation of novel dibenzo[c.e][1,2]oxaphosphorine-based P-ligands. *Transit. Metal Chem.* 2008, 33, 459-465.

[22] Keglevich, G.; Kollár, L. Platinum complexes of five- and six-membered P-heterocycles as potential catalysts. *Lett. Org. Chem.* 2010, 7, 612-620.

[23] Kerényi, A.; Kovács, V.; Körtvélyesi, T.; Ludányi, K.; Drahos, L.; Keglevich, G. A new family of platinum(II) complexes incorporating five- and six-membered cyclic phosphine ligands. *Heteroatom Chem.* 2010, 21, 63-70.

[24] Keglevich, G.; Bagi, P.; Szöllősy, Á.; Körtvélyesi, T.; Pongrácz, P.; Kollár, L.; Drahos, L. Platinum(II) complexes incorporating racemic and optically active 1-alkyl-3-phospholenes and 1-propyl-phospholane P-ligands: Synthesis, stereostructure, NMR properties and catalytic activity. *J.Organomet. Chem.* 2011, 696, 3557-3563.

[25] Gourdel, Y.; Ghanimi, A.; Pellon, P.; Le Corre, M. Activation of phosphines with borane. *Tetrahedron Lett.* 1993, 34, 1011-1012.

[26] Novák, T.; Ujj, V.; Schindler, J.; Czugler, M.; Kubinyi, M.; Mayer, Z. A.; Fogassy, E.; Keglevich, G. Resolution of 1-substituted-3-methyl-3-phospholene 1-oxides by molecular complex formation with TADDOL derivatives. *Tetrahedron: Asymmetry* 2007, 18, 2965-2972.

[27] Ujj, V.; Bagi, P.; Schindler, J.; Madarász, J.; Fogassy, E.; Keglevich, G. A practical and efficient method for the resolution of 3-phospholene 1-oxides via coordination complex formation. *Chirality* 2010, 22, 699-705.

[28] Engel, R. In: Engel, R. ed. *Handbook of Organophosphorus Chemistry*; New York, Marcel Dekker, Inc., 1992, Ch. 5, p. 193.

[29] Pietrusiewicz, K. M.; Flis, A.; Ujj, V.; Körtvélyesi, T.; Drahos, L.; Pongrácz, P.; Kollár, L.; Keglevich, G. Synthesis and use of borane and platinum(II) complexes of 3-diphenylphosphino-1- phenylphospholane (LuPhos). *Heteroatom Chem.* 2011, 22, 730-736.

[30] Keglevich, G. Synthesis of 6-membered and 7-membered P-heterocycles by ring enlargement. *Synthesis* 1993, 931-942.

[31] Keglevich, G.; Böcskei, Z.; Újszászy, K.; Tőke, L. Synthesis of 1,2,3,6-tetrahydrophosphorin 1-oxides by regioselective reduction. *Synthesis* 1997, 1391-1393.

[32] Keglevich, G.; Szöllősy, A.; Tőke, L.; Fülöp, V.; Kalmán, A. Formation of the 4-methylene-1,4-dihydrophosphinine ring-system in the reaction of 2,5-dihydro-3,4-dimethyl-1H-phosphole 1-oxides with dichlorocarbene. *J.Org. Chem.* 1990, 55, 6361-6362.

[33] Keglevich, G.; Sipos, M.; Imre, T.; Ludányi, K.; Szieberth, D.; Tőke, L. Diastereoselective synthesis of 1,2,3,6-tetrahydrophosphinine 1-oxides with an exocyclic P-function by a Michael type addition. *Tetrahedron Lett.* 2002, 43, 8515-8518.

[34] Keglevich, G.; Sipos, M.; Szieberth, D.; Nyulászi, L.; Imre, T.; Ludányi, K.; Tőke, L. Weak intramolecular interactions as controlling factors in the diastereoselective formation of 3-phosphinoxido- and 3-phosphono-1,2,3,6-tetrahydrophosphinine 1-oxides. *Tetrahedron* 2004, 60, 6619-6627.

[35] Czugler, M.; Körtvelyesi, T.; Fábián, L.; Sipos, M.; Keglevich, G. Intra- and intermolecular interactions and water pincer in the crystal structure of a 3-P(O)Ph_2 substituted 1,2,3,6-tetrahydrophosphinine oxide hydrate. *Cryst. Eng. Comm.* 2007, 9, 561-565.

[36] Keglevich, G.; Sipos, M.; Szieberth, D.; Petőcz, G.; Kollár, L. 4-Chloro-5-methyl-3-diphenylphosphino-1-phenyl-1,2,3,6-tetrahydrophosphinine as a bidentate P-ligand in a cis chelate Pt(II) complex. *J. Organomet. Chem.* 2004, 689, 3158-3162.

[37] Keglevich, G.; Sipos, M.; Körtvélyesi, T.; Imre, T.; Tőke, L. 1,2,3,4,5,6-Hexahydrophosphinine 1-oxides with an exocyclic P-function at position 3: diastereoselective synthesis, stereostructure and conformation. *Tetrahedron Lett.* 2005, 46, 1655-1658.

[38] Körtvélyesi, T.; Sipos, M.; Keglevich, G. A theoretical study on the conformation of 3-phosphinoxido- and 3-phosphono-1,2,3,4,5,6-hexahydrophosphinine 1-oxides *Heteroatom Chem.* 2005, 16, 520-524.

[39] Keglevich, G.; Sipos, M.; Ujj, V.; Kötvélyesi, T. Diastereoselective synthesis of 3-phosphinoxido- and 3-phosphono-1,2,3,4,5,6-hexahydrophosphinine oxides as potential precursors of bidentate P-ligands. *Lett. Org. Chem.* 2005, 2, 608-612.

[40] Kukhar, V. P.; Hudson, H. R. eds. *Aminophosphonic and aminophosphinic acids: Chemistry and biological activity.* Chichester, Wiley, 2000.

[41] Fields, S. C. Synthesis Of Natural Products Containing A C-P Bond. *Tetrahedron* 1999, 55, 12237-12273.

[42] Grembecka, J.; Mucha, A.; Cierpicki, T.; Kafarski, P. The most potent organophosphorus inhibitors of leucine aminopeptidase. Structure-based design, chemistry, and activity. *J. Med. Chem.* 2003, 46, 2641-2655.

[43] Kabachnik, M. I.; Medved, T. Y. A new method for the synthesis of α-aminophosphoric acids. *Dokl. Akad. Nauk SSSr*, 1952, 689-692; *Chem. Abstr.*, 1953, *47*, 2724b.

[44] Fields, E. K. The synthesis of esters of substituted amino phosphonic acids. *J. Am. Chem. Soc.*, 1952, 74, 1528-1531.

[45] Cherkasov, R. A.; Galkin, V. I. The Kabachnik–Fields reaction: synthetic potential and the problem of the mechanism. *Russ. Chem. Rev.*, 1998, 67, 857-882.

[46] Matveeva, E. D.; Podrugina, T. A.; Tishkovskaya, E. V.; Tomilova, L. G.; Zefirov, N. S. A novel catalytic three-component synthesis (Kabachnick–Fields reaction) of α-aminophosphonates from ketones. *Synlett*, 2003, 2321-2324.

[47] Ranu, B. C.; Hajra, A. A simple and green procedure for the synthesis of α-aminophosphonate by a one-pot three-component condensation of carbonyl compound, amine and diethyl phosphite without solvent and catalyst. *Green Chem.*, 2002, 4, 551-554.

[48] Zahouily, M.; Elmakssoudi, A.; Mezdar, A.; Rayadh, A.; Sebti, S. Uncatalysed preparation of α-amino phosphonates under solvent free conditions. *J. Chem. Res.*, 2005, 324-327.

[49] Kabachnik, M. M.; Zobnina, E. V.; Beletskaya, I. P. Catalyst-free microwave-assisted synthesis of α-aminophosphonates in a three-component system: (RC)-C-1(O)R-2-(EtO)(2)P(O)H-RNH_2. *Synlett*, 2005, 1393-1396.

[50] Mu, X.-J.; Lei, M.-Y.; Zou, J.-P.; Zhang, W. Microwave-assisted solvent-free and catalyst-free Kabachnik–Fields reactions for α-amino phosphonates. *Tetrahedron Lett.*, 2006, 47, 1125-1127.

[51] Keglevich, G.; Szekrényi, A. Eco-friendly accomplishment of the extended Kabachnik–Fields reaction; a solvent- and catalyst-free microwave-assisted synthesis of α-aminophosphonates and α-aminophosphine oxides. *Lett. Org. Chem.*, 2008, 5, 616-622.

[52] Kiss, N. Z.; Kaszás, A.; Drahos, L.; Mucsi, Z.; Keglevich, G. A neighbouring group effect leading to enhanced nucleophilic substitution of amines at the hindered alpha-carbon atom of an alpha-hydroxyphosphonate. *Tetrahedron Lett.* 2012, 53, 207-209.

[53] Cherkasov, R. A.; Garifzyanov, A. R.; Talan, A. S.; Davletshin, R. R.; Kurnosova, N. V. Synthesis of new liophilic functionalized aminomethylphosphine oxides and their acid-base and membrane-transport properties toward acidic substrates. *Russ. J. Gen. Chem.* 2009, 79, 1480-1494.

[54] Priya, S.; Balakrishna, M. S.; Mague, J. T.; Mobin, S. M. Insertion of carbon fragments into P(III)-N bonds in phosphinoamines and bis(phosphino)amines: Synthesis, reactivity and coordination chemistry of resulting phosphine oxide derivatives. Crystal and molecular structures of $(Ph_2P(O)CH_2)_2NR$ (R=Me, ^{n}Pr, ^{n}Bu), $Ph_2P(O)CH(OH)^n$Pr and *cis* $[MoO_2Cl_2\{(Ph_2P(O)CH_2)_2NEt\text{-}\kappa O,\kappa O\}]$. *Inorg. Chem.* 2003, 42, 1272-1281.

[55] Keglevich, G.; Szekrényi, A.; Szöllősy, Á.; Drahos, L. Synthesis of bis(phosphonatomethyl)-, bis(phosphinatomethyl)-, and bis(phosphinoxidomethyl)-amines, as well as related ring bis(phosphine) platinum complexes. *Synth. Comm.* 2011, 41, 2265-2272.

[56] Bálint, E.; Fazekas, E.; Pintér, G.; Szöllősy, Á.; Holczbauer, T.; Czugler, M.; Drahos, L.; Körtvélyesi, T.; Keglevich, G. Synthesis and utilization of the bis(>$P(O)CH_2$)amine derivatives obtained by the double Kabachnik–Fields reaction with cyclohexylamine; quantum chemical and X-ray study of the related bidentate chelate platinum complexes. *Curr. Org. Chem.* 2011, 16, 547-554.

[57] Bálint, E.; Fazekas, E.; Pongrácz, P.; Kollár, L.; Drahos, L.; Holczbauer, T.; Czugler, M.; Keglevich, G. *N*-Benzyl and *N*-aryl bis(phospha-Mannich adducts): synthesis and catalytic activity of the related bidentate chelate platinum complexes in hydroformylation. *J. Org. Chem.* under publication.

[58] Frisch, M. J.; Trucks, G. W.; Schlegel, H. B.; Scuseria, G. E.; Robb, M. A.; Cheeseman,

J. R.; Montgomery, J. A. Jr.; Vreven, T.; Kudin, K. N.; Burant, J. C.; Millam, J. M.; Iyengar, S. S.; Tomasi, J.; Barone, V.; Mennucci, B.; Cossi, M.; Scalmani, G.; Rega, N.; Petersson, G. A.; Nakatsuji, H.; Hada, M.; Ehara, M.; Toyota, K.; Fukuda, R.; Hasegawa, J.; Ishida, M.; Nakajima, T.; Honda, Y.; Kitao, O.; Nakai, H.; Klene, M.; Li, X.; Knox, J. E.; Hratchian, H. P.; Cross, J. B.; Adamo, C.; Jaramillo, J.; Gomperts, R.; Stratmann, R. E.; Yazyev, O.; Austin, A. J.; Cammi, R.; Pomelli, C.; Ochterski, J. W.; Ayala, P. Y.; Morokuma, K.; Voth, G. A.; Salvador, P.; Dannenberg, J. J.; Zakrzewski, V. G.; Dapprich, S.; Daniels, A. D.; Strain, M. C.; Farkas, O.; Malick, D. K.; Rabuck, A. D.; Raghavachari, K.; Foresman, J. B.; Ortiz, J. V.; Cui, Q.; Baboul, A. G.; Clifford, S.; Cioslowski, J.; Stefanov, B. B.; Liu, G.; Liashenko, A.; Piskorz, P.; Komaromi, I.; Martin, R. L.; Fox, D. J.; Keith, T.; Al-Laham, M. A.; Peng, C. Y.; Nanayakkara, A.; Challacombe, M.; Gill, P. M. W.; Johnson, B.; Chen, W.; Wong, M. W.; Gonzalez, C.; Pople, J. A. *Gaussian 03, Revision B.05*, Gaussian, Inc., Pittsburgh PA, 2003.

[59] PCMODEL, Version 7.0, 2005, Bloomington, IN, USA.

[60] Stewart, J. J. P Optimization of Parameters for Semiempirical Methods V: Modification of NDDO Approximations and Application to 70 Elements *J. Mol. Mod.* 2007, 13, 1173-1213.

[61] Stewart, J.J.P. MOPAC2007 and MOPAC2009, www.openmopac.net, USA.

[62] Rezac, J.; Hobza, P. Advanced Corrections of Hydrogen Bonding and Dispersion for Semiempirical Quantum Mechanical Methods *J. Chem. Theor. & Comp.* 2012, 8, 141-151.

In: Platinum: Compounds, Production and Applications
Editors: L.Varennikov and E. Yedemsky
ISBN: 978-1-62257-939-6

Chapter 4

PLATINUM FILMS ON YSZ WITH DIFFERENT TYPES OF GRAIN BOUNDARIES – MICROSTRUTURE CHANGES DUE TO ANNEALING AND OXYGEN EXCHANGE

G. Beck*

Research Institute Precious Metals and Metals Chemistry, Germany

ABSTRACT

Platinum films were prepared on (111) and (100) orientated and on polycrystalline yttrium-stabilized zirconia (YSZ) by pulsed laser deposition (PLD). Afterwards, the Pt/YSZ systems were first annealed and then electrically polarised. The polarisation was performed in the manner that oxygen was alternatingly built-out and built-in (=oxygen exchange) at the Pt/YSZ. The as-deposited, annealed and polarised films, respectively, were characterised by EBSD, SEM, X-ray pattern and pole figure measurements. The as-deposited platinum films on all used substrates are (111) orientated, but the microstructures are different and depend strongly on the microstructure of the substrate. In as-deposited platinum films on (111) orientated YSZ the only defects are twins of a size between 100 nm and 300 nm. These twins disappear completely during annealing in platinum films on perfect (111) orientated YSZ single crystals and holes are built instead. Whereas on twin-rich (111) YSZ substrates at least a small fraction of twins remain. Nevertheless, during polarisation these twins also disappear completely, due to formation and cracking of oxygen bubbles at twin positions. The platinum films on (100) orientated YSZ are polycrystalline with mostly 30° and 60° grain boundaries. On polycrystalline YSZ the platinum films are formed with all kinds of grain boundary angles. Due to annealing the texture of the platinum films on (100) and polycrystalline YSZ becomes sharper, since the surface becomes smooth. Due to polarisation the grain boundaries are widened and the texture of the platinum films becomes softer, since the surface becomes waved. It can be concluded that twin grains within platinum films behave differently during annealing and oxygen exchange when compared to other grains. Presumably, this

* Correspondence author: phone: +49 7171 1006402, fax: +49 7171 1006900, E-mail: beck@fem-online.de

results from the much smaller disturbed regions within the crystal lattice at the twin grain boundaries.

1. INTRODUCTION

Platinum is an electron conductor, has a high catalytic activity and is even at high temperatures stable against oxidation. Due to these properties platinum films on the oxygen-ion-conductor YSZ (= yttrium-stabilized zirconia) can be used for oxygen built-in or built-out (= oxygen exchange) in the electrochemical cell $Pt(O_2)/YSZ/Pt(O_2)$, by polarisation of the cell. Accordingly, this electrochemical cell is used in oxygen-sensors and in catalytic processes as well as a model system for a solid-oxide-fuel-cell. Certainly, the oxygen exchange occurs primarily at the triple phase boundary between platinum, YSZ and the gas phase [1-7], but it is supposable, a limited oxygen exchange also takes place at lattice defects like grain boundaries or dislocations within the platinum film (since at these defects the crystal lattice is widened) or at the platinum/YSZ interface, e.g. as a thin platinum oxide film (Figure 1).

In the literature there are a few investigations about oxygen exchange at such lattice defects: It had been shown that grain boundaries within the platinum film are oxygen permeable [5] and that the Pt/YSZ interface is a limited oxygen reservoir, where a small amount of oxygen can be chemisorbed or stored in oxygen-filled voids or form a thin platinum oxide layer [6, 7]. Up to now oxygen exchange at lattice defects has not been studied in detail, since the preparation of platinum films on YSZ with a well-defined microstructure is not easy.

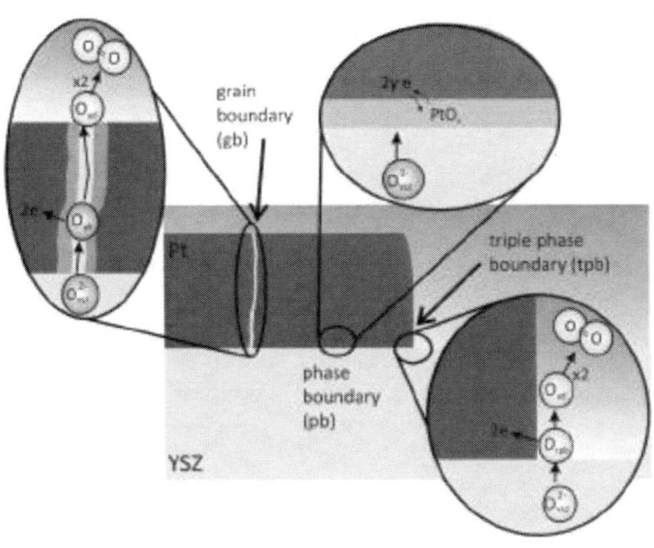

Figure 1. Supposable oxygen built-out reactions in the system Pt/YSZ at the triple-phase boundary (tpb), a grain boundary (gb) and at the interface.

The standard preparation of platinum electrodes on YSZ is based on the sintering of platinum paste, i.e. a dispersion of platinum microcrystals in an organic solvent. Other platinum films have been deposited by sputtering, clcctron beam evaporation and thermal evaporation [8-10]. All these preparation methods produce polycrystalline platinum films. The platinum films formed by sintering of a paste are also very porous. Nearly dense and microstructural well-defined platinum films can be prepared by pulsed laser deposition (PLD) on YSZ substrates and subsequent annealing [11]. The microstructure of the platinum films depends on the microstructure of the substrate, e.g. on (111) orientated twin-rich YSZ these twins can also be found within the (apart from that single-crystalline) platinum films [12]. The pulsed-laser deposited platinum films are not absolutely dense after annealing, since de-wetting phenomena - in which holes are formed within the platinum - occurs during the annealing procedure. The number of holes is higher, when the roughness of the substrateis larger [13]. Therefore, unevenness at the Pt/YSZ interface is supposed to be a location for de-wetting. Other locations for hole formation during de-wetting are droplets on the platinum films, which are built during PLD [12]. Moreover, within such nearly dense annealed platinum films on YSZ the formation and cracking of oxygen bubbles – and accordingly hole formation - during anodic polarisation was found [14, 15]. Up to now, neither in the case of de-wetting nor in the case of polarisation the influence of grain boundaries on the hole formation process has been studied. Therefore, in the following, platinum films on YSZ with different types of grain boundaries were prepared by PLD and the behaviour of the different grain boundaries during annealing and during polarisation was investigated.

2. Experimental

2.1. Preparation of the Platinum Films

The platinum films were prepared by pulsed laser deposition (PLD) with a KrFlaser ($\lambda = 248$ nm) and Argon as background gas (2 Pa, purity 99.95%). The laser had a repetition rate of 6 Hz and pulse energy of 450 mJ. The cylindrical and polished platinum target (Ögussa Austria, purity 99.95 %) has a diameter of 1 cm and is placed at a distance of 4 cm to the substrate in the PLD chamber. The substrates were commercially polished (111) orientated YSZ single crystals (some with a small fraction of twins) as well as (100) orientated and polycrystalline YSZ. The substrate temperature was about 400 °C. The growth rate of the films was about 1 μm/h and films of about 500 nm thickness were prepared.

2.2. Annealing

The as-deposited platinum films were annealed in air at a temperature of 750 °C for 48 hours with a heating and cooling rate of approximately 2 °C/min.

2.3. Polarisation

The polarisations were performed with the electrochemical cell Pt(O_2)/YSZ/Pt-paste(O_2) (Figure 2) in a cyclovoltammetry experiment with a VersaSTAR 3 measurement device

(Princeton Applied Research). Accordingly, the $Pt(O_2)/YSZ$ side was alternatingly polarised as anode or as cathode and analogous oxygen was built-out and built-in there. The experiments were conducted at 400 °C and in air ($p(O_2)$=0,21 bar).

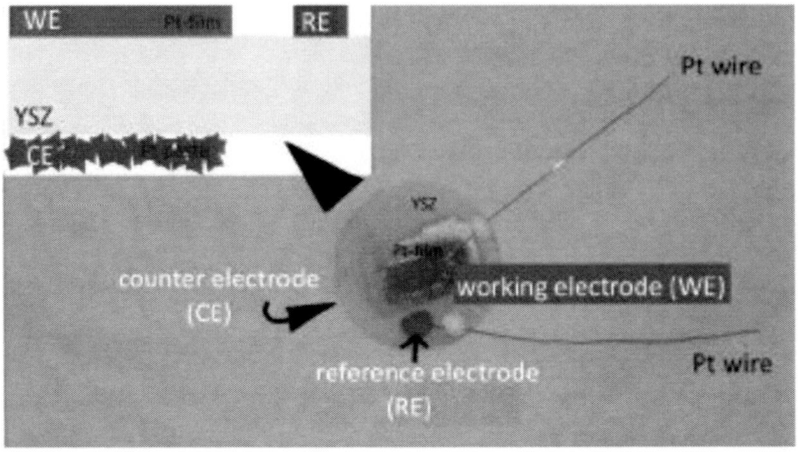

Figure 2. Photo of the electrochemical cell with the platinum film, which is to be investigated (=working electrode) on the transparent YSZ. Inset: Sketch of the cross section.

For each sample 10 cycles were measured in the voltage range between 500 mV (= anodic case) and -1000 mV (= cathodic case) with a rate of 10 mV/s.

2.4. Characterisation

The morphology of the metal films was investigated by scanning electron microscopy (SEM; field emission microscope: Zeiss, Supra 55 VP). For texture investigations we used electron backscatter diffraction (EBSD; additional equipment for the SEM microscope of Crystal, Oxfords Instruments) as well as X-ray scans and pole figure measurements. The X-ray scans and pole figures weremeasured by the D5000 diffractometer (Siemens, Germany). This diffractometer is equipped with an Eulerian cradle and an X-ray lens. We measured the (111) pole figures of the platinum films and of the substrate through the metal films.

3. Results and Discussion

3.1. Effect of Annealing

Figure 3 showsa SEM image of an as-deposited platinum film on (111) orientated YSZ.

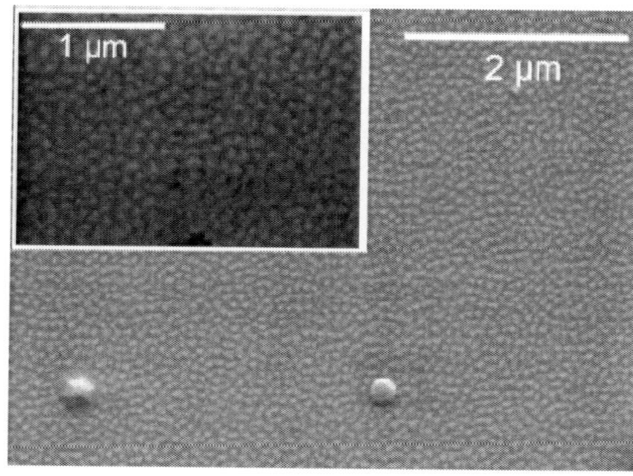

Figure 3. The SEM image of an as-deposited platinum film on (111) YSZ. The typical morphology of a columnar growing process with column heads can be seen. The heads are pyramidal and have sizes between 30 nm and 90 nm.

The as-deposited films on all used substrates look nearly the same. They show a sub-structure of small particles with sizes between 30 nm and 90 nm. These particles are presumably column heads, due to the columnar growth. In addition, droplets (as a result of the deposition process) of µm-size can be found, as also described elsewhere [12]. Figure 4 show the SEM images of annealed platinum films on (111) and on (100) orientated YSZ. In the first case no grains are visible and these films look flat with only some µm-sized triangular or hexagonally shaped holes and droplets (beside a few scratches probably from the handling with the tweezers). One of such triangular shaped droplets is visible in figure 4a and one hexagonally shaped hole can be seen in the inset there. In contrast, the platinum films on (100) (Figure 4b) and also on polycrystalline YSZ (not shown) show a grain structure. On (100) orientated YSZ the grains have sizes between 500 nm and 10,000 nm and on polycrystalline YSZ the grains are a bit smaller with sizes between 400 nm and 2,000 nm. Also, in these polycrystalline platinum films, µm-sized holes and droplets are visible, but they do not have such a pronounced triangular or hexagonal shape and the number of holes is much smaller than in the case of the annealed films on (111) YSZ. The X-ray scans (not shown) show that the as-deposited and annealed platinum films are always (111) orientated independent of the used substrate, obviously as a result of the smallest surface energy of the (111) surface in the face-centred cubic crystals platinum [12].

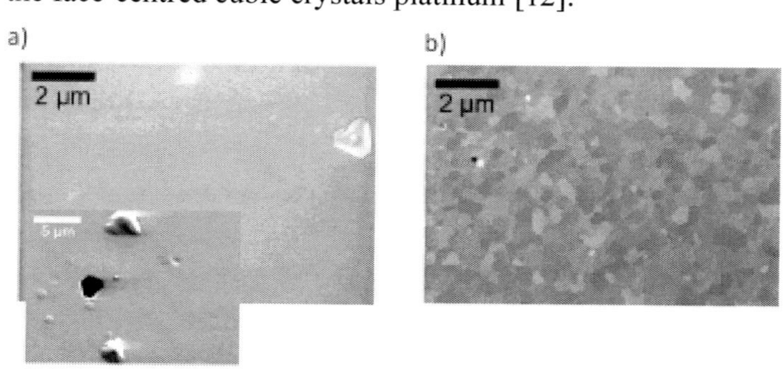

Figure 4. The SEM image of an annealed platinum film on a) (111) and b) (100) orientated YSZ. In the first case no grain structure is visible, but triangular and hexagonal holes and droplets. In the second case a grain structure can be seen.

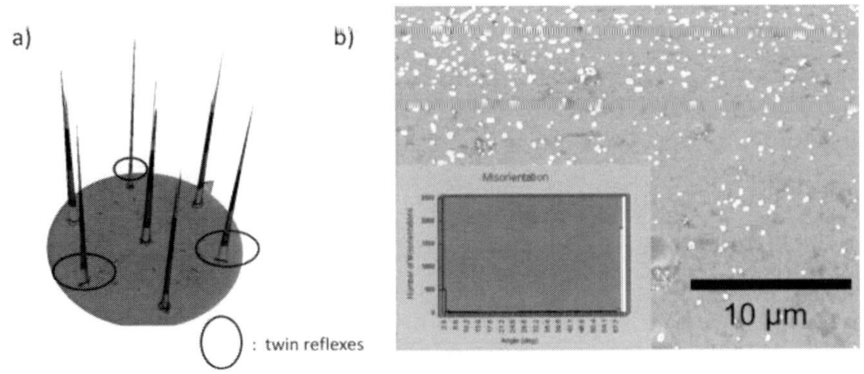

Figure 5. Three-dimensional plot of the {111} pole figure (a) and the grain boundary (or misorientation) map (b) of the as-prepared platinum-film on (111) orientated YSZ. A large number of twins can be seen. The number of misorientations with the twin angle of 60° is approximately 2500 within the investigated area.

The texture investigations clarify the detailed microstructure. In the case of the (111) orientated YSZ as substrate the as-deposited platinum films have always a fraction of twins (Figure 5). The EBSD images show, these twins have sizes between 100 nm and 300 nm. These twins can disappear completely in the case of a perfect (111) single crystal as substrate (Figure 6a). If the substrate is twin-rich a fraction of twins remains within the platinum film after annealing (Figure 6b and 6c). Figures 6b and 6c show that the fraction of twins can be very different, although the fraction of twins within the substrate was comparable here. The reasons for this are not understood up to know.

The EBSD investigations confirm these results: the twin fraction is decreased or removed completely. Figure 7a shows a grain boundary map determined by EBSD of a platinum film on a perfect (111) YSZ substrate - with no twins (compare Figure 6a). Here only some defects (= small angle grain boundaries) at scratches within the platinum film are visible. In the grain boundary map of a platinum film on twin-rich YSZ the remained twins (Figure 7b) can be seen as white lines. The inset in this figure shows the distribution of the grain boundaries (from 0° up to about 60°) of the investigated area.

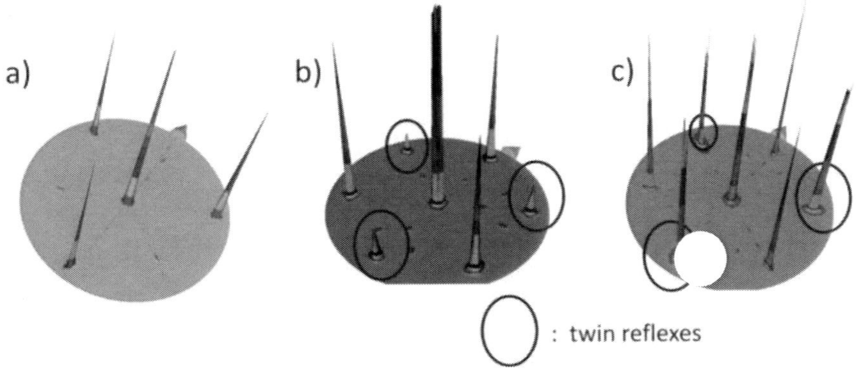

Figure 6. Three-dimensional plot of the {111} pole figure of platinum of annealed platinum films on (111) orientated YSZ. Different concentrations of twins - increasing from a) up to b) up to c) – can be seen.

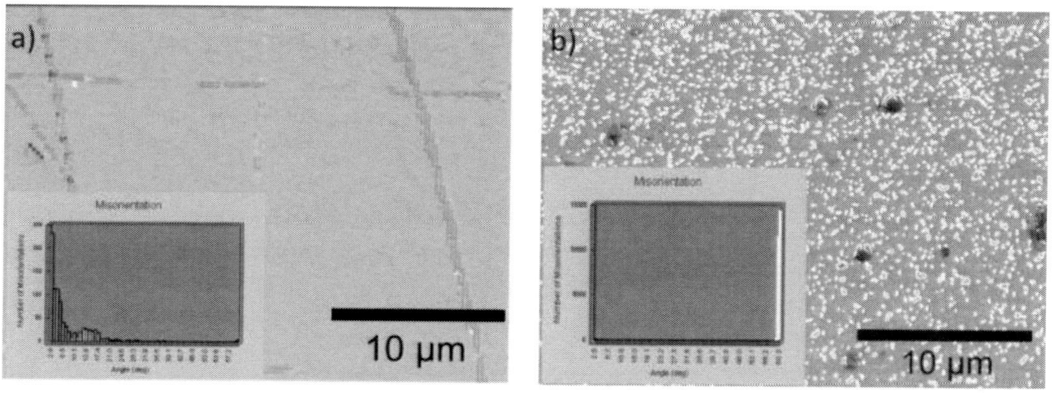

Figure 7. Grain boundary maps of annealed platinum films on YSZ (111) a) with no twins and b) with a large number of twins. Insets: Corresponding grain boundary angle distribution. In a) an area of the platinum film is shown, where only a few small angle grain boundaries were formed at scratches within the film.In b) a large number of twins are visible. The number of misorientations with the twin angle of 60° is here approximately 15000.

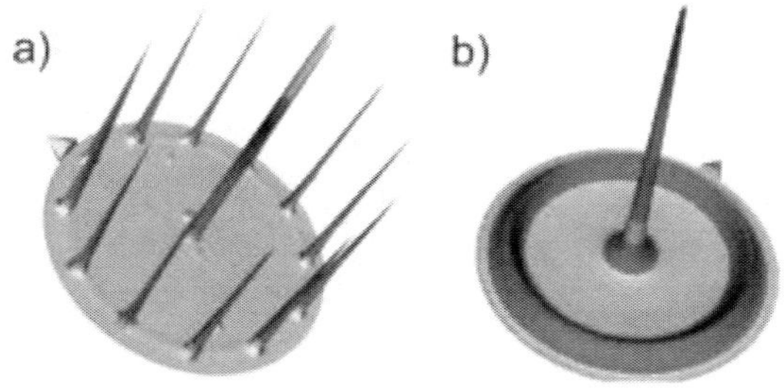

Figure 8. Three-dimensional plot of the {111} pole figure of platinum of annealed platinum films on a) (100) orientated and b) polycrystalline YSZ.

Only twin grain boundaries (60° angle grain boundaries) are found. It should be remarked that the twins are not visible in the SEM investigations, but only in the EBSD and X-ray pole figure investigations. In figure 8 the pole figures of the annealed platinum films on (100) orientated and on polycrystalline YSZ are shown. On (100) YSZ 12 peaks with the maxima at a tilt angle of 70.5° occur (with a rotation angle distance of 30°). Accordingly, there are four energetically equivalent positions for platinum on this substrate - corresponding to the fourfold symmetry of the substrate [12]. On the polycrystalline substrate a “reflex-ring” with its maximum at 70.5° can be seen, i.e. this platinum film has a fibre texture. Accordingly, there are macroscopically no energetically favoured positions identified for the growth of the platinum grains on this substrate.

Figure 9 shows the rotation scan at a tilt angle of 70.5° before and after annealing. It can be seen that due to annealing the reflexes become smaller in the rotation direction. Not shown is that the reflexes in the tilt direction also become smaller due to annealing. In the case of platinum on polycrystalline YSZ the ring also becomes smaller in the tilt direction. Therefore, in both cases due to annealing the reflexes became smaller and correspondingly the texture became sharper, but the characteristics of the pole figures remain. Figure 10 shows the corresponding EBSD images of the annealed platinum films on (100) and polycrystalline YSZ.

The insets show the angles between the grain boundaries. In the case of platinum on (100) YSZ mainly 30° and 60° grain boundary angles can be found. This is consistent with the 30° rotation angle distance, which had been found in the pole figure (Figure 8a).Accordingly, here are a lot of the twin grain boundaries (60° angles) and (as mentioned above) not visible in the SEM images.

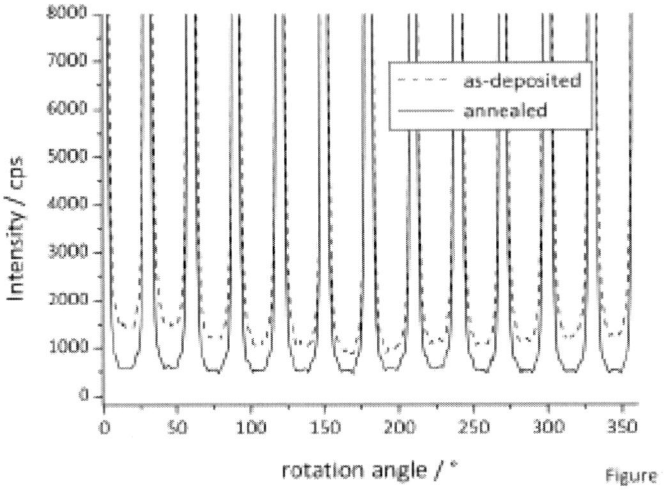

Figure 9. X-ray rotation scan at a tilt angle of 70.5° of the platinum film on (100) orientated YSZ before and after annealing.

In the case of platinum on polycrystalline YSZ all kind of grain boundary angles can be found with a small preference for small angles and twin grain boundaries.

3.2. Effect of Oxygen Exchange

The platinum films with different fractions of twins (compare Figure 6) and the polycrystalline platinum films (compare figures. 8 and 10) have been polarised within the cyclovoltammetry experiment (at 400 °C in air). Figures 11 and 12 clarify the twin behaviour during oxygen exchange: Here the SEM images and the corresponding pole figures of a platinum film with only a few twins (Figure 11) and of a platinum film with several twins (Figure 12), respectively, before and after polarisation can be seen. In both cases the twins have disappeared completely after polarisation and holes have formed instead. In the case of a small fraction of twins within the platinum film only a few - but large - holes are formed.

In the case of a large fraction of twins a large number of small holes are formed. Also the grain boundary maps (insets in Figure 12) prove the disappearance of the twins. These results indicate that the bubble formation and cracking during the anodic process (i.e. the oxygen built-out) takes place at twin positions, if no other grain boundaries are within the platinum film.

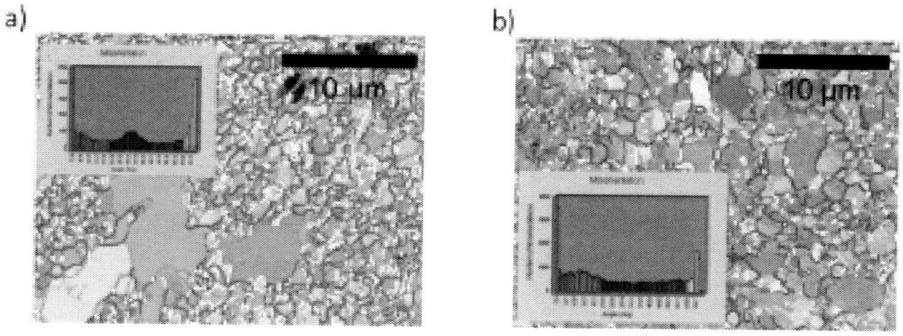

Figure 10. Grain boundary maps of annealed platinum films on a) (100) orientated and b) polycrystalline YSZ. Insets: Corresponding grain boundary angle distributions. In a) a maximum around 30° and 60° grain boundary angles is evident. In b) the grain boundary angles are nearly homogeneously distributed with a small preference for small angle and twin grain boundaries.

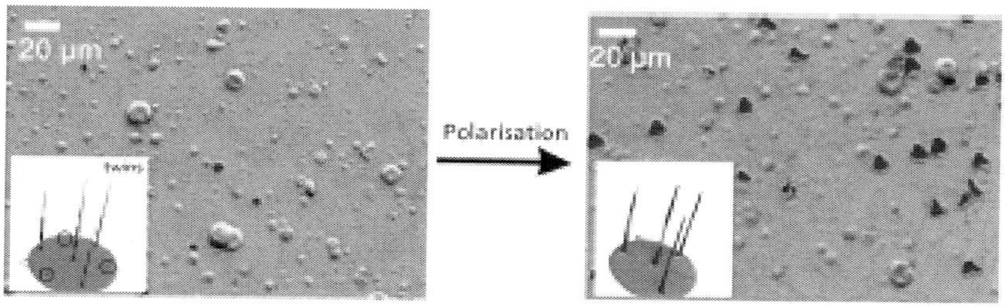

Figure 11. Annealed platinum film on (111) orientated twin-rich YSZ before polarisation (left) and afterwards (right). Shown are the SEM images and as insets the {111} pole figures. The platinum film before polarisation has a small number of holes and a relatively large number of droplets as well as a small number of twins. Due to polarisation several large holes have been formed within the film and the twins have disappeared.

Figure 13 illustrates the same investigations for the polycrystalline platinum film on (100) YSZ. Hole formation is not visible in the SEM investigations, but the metal film is waved and the grain boundaries are widened after polarisation.

The pole figures show no significant difference due to polarisation, only a broadening of the reflexes, probably a result of the waved metal surface. In the case of the polycrystalline platinum film on polycrystalline YSZ in principle the same results are found: the grain boundaries are widened and the reflex-ring is broadened due to polarisation.

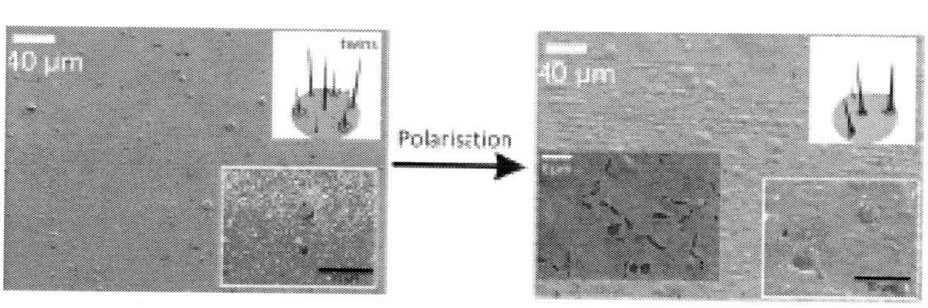

Figure 12. Annealed platinum film on (111) orientated twin-rich YSZ before polarisation (left) and afterwards (right). Shown are the SEM images and as insets the {111} pole figures (top) as well as the grain boundary maps (bottom). The platinum film before polarisation has a small number of holes and droplets as well as a large number of twins. Due to polarisation a large number of small holes have been formed within the film and the twins have disappeared.

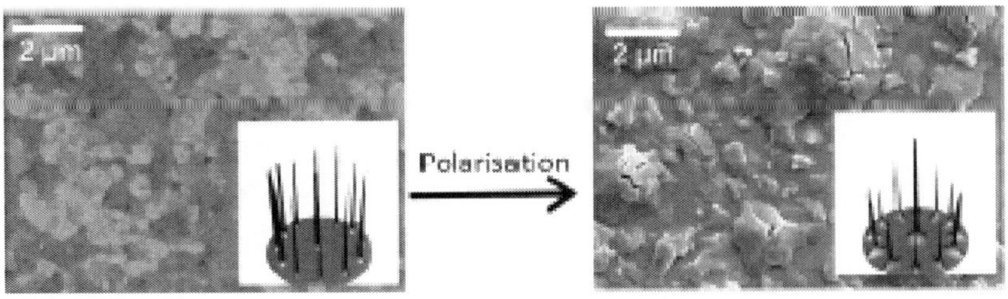

Figure 13. Annealed platinum film on (100) orientated YSZ before polarisation (left) and afterwards (right). Shown are the SEM images and as insets the {111} pole figures (bottom). The platinum film shows 12 reflexes (indicating 30° and 60° grain boundaries) before and after polarisation. Due to annealing the reflexes are broadened, probably as a result of the waved surface, which is visible in the SEM images.

3.3. Summary

By pulsed laser deposition and sub-sequentially annealing platinum films with different types of grain boundaries were prepared: On single crystalline (111) orientated YSZ the platinum films are single crystalline, but on twin-rich (111) orientated YSZ twins (grains with 60° grain boundaries) can be found as defects. Platinum films on (100) orientated YSZ have grains mostly with 30° and 60° grain boundaries, whereas all types of grain boundary angles can be found between platinum grains on polycrystalline YSZ (with a small preference for small angles and twin grain boundaries). With these platinum films the behaviour of the different grain boundaries during annealing and during oxygen exchange were investigated. The investigations show that 60° twin grain boundaries behave differently from other grain boundaries (the 30° grain boundaries show no special behaviour). The difference is that the whole twin grains are removed after annealing or after oxygen exhange and holes are formed instead, whereas in the case of the other grain boundaries the platinum film is smoothed after annealing and the grain boundaries are widened after oxygen exchange, respectively.

Twin boundaries are special types of high-angle grain boundaries and twins can be described with a twinning-plane between the twin grains. Twinning-planes can be arranged in different ways within a crystal. In the investigated platinum films the twinning-planes are perpendicular to the surface. Accordingly, these twins are called contact twins. They can also be described by a 60° rotation around the surface normal (compare pole figure measurements Figures 5a, 6b and 6c) and have a 60° grain boundary angle (compare EBSD investigations, Figures 5b and 7b). In the ideal case twin boundary energies (about 10 to 30 mJ/m^2) are much smaller than that of other high-angle grain boundaries (about 400 to 900 mJ/m^2) and even smaller than that of small-angle grain boundaries (about 250 mJ/m^2) [16,17]. The ideal case is a coherent twin boundary. Here the boundary is parallel to the twinning plane. If the twin boundary is not exactly parallel to the twinning plane, the boundary energies are higher (about 100 to 500 mJ/m^2). This is known as an incoherent twin boundary. Between small-angle and high-angle grain boundaries disturbed regions exists in the crystal lattice and correspondingly the lattice is widened here – in the case of a small-angel grain boundary the disturbed region is smaller than in the case of a high-angle grain boundary. At a coherent twin boundary the lattice is not widened, because the atoms in the boundary are essentially in undistorted positions. At a semicoherent or incoherent twin boundary the atoms do not fit perfectly into each grain and therefore, the lattice is here also widened a bit. The nearly round twins – which are visible in figures 5b and 7b - cannot be separated only by coherent twin boundaries. Therefore, also semicoherent or incoherent twin boundaries have to exist here and the platinum lattice is widened here. Obviously, these widened lattice is the location for hole formation during annealing and polarisation.

The formed holes are µm-sized, whereas the size of the twin grains is much smaller (between 100 nm and 300 nm, compare Figures 5b and 7b). Accordingly, the twin grains are only the starting point for the hole formation. In the case of annealing, de-wetting phenomenon at such starting defects are known [18]. It can be assumed that within the as-deposited columnar structure strain is incorporated, which relaxes due to the hole formation during annealing. In the case of polarisation it is supposable that the oxygen bubbles are formed at the interface Pt/YSZ at the twin positions. When the pressure of these bubbles is high enough they crack and holes are formed in their place. Comparing figure 11 and 12 clarify that at a few twins' big holes – and accordingly big bubbles – and at a large number of twins' small holes – and accordingly small bubbles – are formed. An obvious question is, what happened to the platinum where the hole was formed: has it incorporated within the metal film or has it formed a zone at the hole edge. In the case of annealing, the edges of the holes look sharp (inset Figure 4a) - in the case of bubble formation and cracking, the hole edges look more lacerated (especially Figure 12). Therefore, during annealing the material of the holes is probably incorporated within the metal film and during oxygen exchange it forms a special zone at the hole edges. The observations for the polycrystalline platinum films on (100) and polycrystalline YSZ – i.e. that the films are smoothed after annealing and the grain boundaries are widened after polarisation, respectively – are surely a result of the larger disturbed regions with a wider crystal lattice at the here available low-angle and high-angle grain boundaries. During annealing it is supposable that the inner strain of the as-deposited columnar structure can relax at these grain boundaries and the whole film can become smooth. During polarisation the oxygen exchange can take place within the wide lattice at the grain boundaries. Obviously here, oxygen bubbles are also formed, which press the grain boundaries apart and, accordingly, these platinum films become waved during polarisation.

Conclusion

On the basis of the investigations it can be concluded that twin grains within platinum films behave differently during annealing and polarisation when compared to other grains. It can be assumed that during annealing strain relaxes within the as-deposited platinum films with its columnar structure. If there are only twin grains within the platinum film, holes are formed at the twin positions for this relaxation. If there are grains with other grain boundaries there are wider disturbed regions at these grain boundaries, which allow the relaxation of the strain. During polarisation oxygen is exchanged mostly at the triple phase boundary, but obviously also at twin positions and at the grain boundaries of other grains. In the case of twin grains only small disturbed regions are at the twin boundaries in the crystal lattice (and only at semi- or incoherent twin boundaries, but not at coherent twin boundaries). These small disturbed regions are obviously too small for oxygen exchange and, therefore, oxygen bubbles are formed here and crack – finally forming holes at the twin positions. In the case of grains with other grain boundary angles, larger disturbed regions exist in the crystal lattice, allowing oxygen exchange and these boundaries are only widened due to polarisation.

Acknowledgments

Financial support by the DFG (Deutsche Forschungsgemeinschaft) in the projects BE 3170/3-1 is gratefully acknowledged. I also thank H. Pöpken (Justus-Liebig-University Giessen, Germany) for the preparation of the platinum films and J.W. Ashford for revising the manuscript.

References

[1] J. Mizusaki, K. Amano, S. Yamauchi, K. Fueki, *Solid State Ionics* 22(4) (1987) 313.
[2] A. Mitterdorfer, L. J. Gauckler, *Solid State Ionics* 117(3,4) (1999) 203.
[3] T. M. Gur, R. A. Huggins, *J. Appl. Electrochem.* 17(4) (1987) 800.
[4] S. B. Adler, *Chem. Rev.* 104 (2004) 4791.
[5] T. Ryll, H. Galinski, L. Schlagenhauf, P. Elser, J.L.M. Rupp, A. Biberle-Hutter, L.J. Gauckler, *Adv. Funct. Mater.*21 (2011) 565–572.
[6] A.K. Opitz, J. Fleig, *Solid State Ionics* 181 (2010) 684-693.
[7] G. Foti, A. Jaccoud, C. Falgairette, C. Comninellis, J. Electroceram23 (2009)175-79.
[8] J. E. Bauerle, J. Phys. Chem. Sol. 30 (1969) 2657.
[9] R. J. Brook, W. L. Pelzmann, and F. A. Kröger, *J. Electrochem. Soc.* 118(2) (1971) 185.
[10] K. Sakurai, H. Nagamoto, and H. Inoue, *Solid State Ionics* 35 (1989) 405.
[11] G. Beck, H. Fischer, E. Mutoro, V. Srot, K. Petrikowski, E. Tchernychova, M. Wuttig, M. Rühle, B. Luerßen, J. Janek, *Solid State Ionics* 178 (2007) 327.
[12] G. Beck, H. Pöpke, B. Luerßen, J. Janek, *J. Cryst. Growth* 322 (2011) 95–102.
[13] N. Baumann, E. Mutoro, J. Janek, *Solid State Ionics* 181 (2010) 7.
[14] E. Mutoro, S. Günther, B. Luerßen, I. Valov, J. Janek, *Solid State Ionics* 179 (2008) 1835.
[15] H. Pöpke, E. Mutoro, B. Luerßen, J. Janek, *Solid State Ionics* 189 (2011) 56.
[16] D.A. Porter, K.E. Easterling, *Phase Transformation in Metals and Alloys*, Second Edition, ChapmanandHalll London, 1992.
[17] H.-J. Bargel, G. Schulze, 10. Edition 2008, *Springer Verlag Berlin Heidelberg*, Werkstoffkunde.
[18] E. Shaffir, I. Riess, W.D. Kaplan, *Acta Mater.* 57 (2009) 248.

In: Platinum: Compounds, Production and Applications
Editors: L.Varennikov and E. Yedemsky
ISBN: 978-1-62257-939-6

Chapter 5

PLATINUM NANOPARTICLES AND INFLAMMATION

Yoko Yoshihisa and Tadamichi Shimizu*
Department of Dermatology, Graduate School of Medicine,
University of Toyama, Sugitani, Toyama, Japan

ABSTRACT

Platinum nanoparticles (nano-Pts) possess anti-inflammatory activity and the ability to scavenge superoxides anions and peroxides, indicating that they can act as superoxide dismutase (SOD)/catalase mimetics. These potentials seem useful in the protection and/or the amelioration of oxidative stress-associated pathologies. Recent studies investigated the effects of poly acrylic acid-capped nano-Pts on ultraviolet (UV)-induced inflammation and the underlying molecular mechanisms in keratinocytes. Nano-Pts effectively protect against UV-induced inflammation in keratinocytes by decreasing reactive oxygen species (ROS) production. In addition, nano-Pts significantly inhibit RANKL-stimulated osteoclastogenesis in RAW 264.7 cells by the suppression of ROS. Moreover, the intranasal administration of nano-Pts prior to cigarette smoke exposure has been shown to inhibit inflammation in the lungs of mice. On the other hand, nano-Pts resulted in hyperthermia-desensitization in lymphoma cells where the signaling pathways involved in apoptosis are inhibited by nano-Pts. This chapter reviews the latest findings regarding the effects of nano-Pts on inflammatory responses.

Keywords: Platinum nanoparticles, heat, ultraviolet, apoptosis, ROS, inflammation

1. INTRODUCTION

Nanotechnology is becoming increasingly important for research on the development of new therapeutic approaches for cancer, inflammation and aging. Nanoparticles are currently being applied not only in material science and engineering, but also in medical science and

* Corresponding author.Tadamichi Shimizu, M.D., Ph.D. Department of Dermatology, Graduate School of Medicine and Pharmaceutical Sciences, Tel: +81-76-434-7305; Fax: +81-76-434-5028; E-mail: shimizut@med.u-toyama.ac.jp.

for clinical use. Nanoparticles of some noble metals, including platinum (Pt), act as reducing catalysts due to their large surface area [1]. Platinum nanoparticles (nano-Pts) protected by polyacrylic acid (PAA) were recently manufactured through reduction with ethanol for 2 h [2]. These nano-Pts scavenge superoxide anions ($O_2^{\cdot -}$) and peroxides (H_2O_2) indicating that they can act as superoxide dismutase (SOD)/catalase mimetics [2]. SOD/catalase mimetics can be co-administered with certain therapeutics to minimize reactive oxygen species (ROS)-induced damage to normal tissues and thus alleviate some of their side effects [3]. Also, the treatment of alveolar-like A549 cells with nano-Pts inhibited cell death after exposure to cigarette smoke extract. These results suggested that nano-Pts can act as antioxidants inhibiting the pulmonary inflammation induced by acute cigarette smoking [4]. On the other hand, another study showed that nano-Pts are capable of inducing DNA damage and p53-mediated growth arrest [5]. Moreover, FePt@CoS2 yolkshell nano-Pts were more potent in killing HeLa cells in comparison to cisplatin treatment, suggesting their role as anticancer agents [6]. Therefore, nano-Pts in a biological system can exert a number of effects with positive impact in cancer prevention and treatment. However, the knowledge that the cell killing induced by some of the available therapeutic agents is mediated by an elevation in intracellular ROS raises skepticism about the use of nanoparticles in combination with other modalities. This wary extrapolates from the fact that the addition of exogenous free radical scavengers is widely accepted as an ordinary laboratory maneuver to confirm the role of ROS in apoptosis following various treatments, which results in a proportional increase in the final cell viability. Similarly, nano-Pts, which are potent free radical scavengers and SOD/catalase mimetics, may act to reverse apoptosis which would be detrimental to adjuvant cancer therapy [2].

2. Ultraviolet-Induced Skin Inflammation

UV radiation is divided into 3 subtypes, UVA (320–400nm), UVB (280–320nm) and UVC (200–280nm), each of which has distinct biological effects. Although UVC is blocked by stratospheric ozone, UVA (90–99%) and UVB (1–10%) reach the surface of the earth [7]. The health risks associated with ozone depletion are caused by enhanced UVA radiation in the environment and increased penetration of UVB [8]. Skin is a major target for toxic insults by UV radiation that is capable of altering its structure and function. Both UVA and UVB radiation generate ROS, such as singlet oxygen, $O_2^{\cdot -}$ radicals, hydroxyl radicals, and H_2O_2, in a variety of cells [9]. These intracellular ROS are thought to play an important role in UV-induced inflammatory responses in the skin resulting in immediate consequences, such as erythema and pigmentation, as well as delayed changes, such as cellular DNA mutations, hyperplasia, and carcinogenesis [10]. Skin responds to UV with a complex defense system of enzymatic and non-enzymatic components to counter the adverse biological effects of ROS. ROS are reduced by antioxidant enzymes, such as SOD, catalase, and glutathione peroxidase, as well as by endogenous and exogenous small molecules, including glutathione and vitamins C and E. Oxidized biomolecules are repaired or replaced by biological protective systems. Nevertheless, the biomolecules are gradually irreversibly oxidized over time, thereby impairing the biological functions in the skin, and eventually leading to aging and age-related diseases [11, 12].

3. Protective Effect of Nano-Pts against Ultraviolet-Induced Skin Inflammation

UVB affects the skin by inducing immunosuppression, premature skin aging, inflammation, and cell death. Sunscreen use is a widely accepted method for primary prevention against sunburn and photoaging. The antioxidant compounds used in these products seem to function as protective substances against UV-induced ROS production in the skin.

Topical antioxidants, such as flavonoids, quercetin, ascorbic acid and β- carotene, diminish UV radiation-mediated oxidative damage to the skin [13-16]. For example, ascorbic acid eliminates most ROS due to the oxidation of ascorbate to monodehydroascorbate and then to dehydroascorbate and has diverse functions to maintain the normal physiological status in humans [15]. In addition, β- carotene and astaxanthin, which are components of carotenoids possess the ability to quench $O_2^{\cdot -}$. The mechanisms underlying the protective effects of carotenoids have been studied in a model of UVA-irradiated human dermal fibroblasts, thus revealing that including ROS generation decreases the antioxidant enzyme activities, promotes membrane perturbation, and also induces the expression of heme oxygenase-1 [16].

Similar outcomes can be also brought about by using nano-sized particles in the form of titanium dioxide (TiO_2) and zinc oxide (ZnO) that can reflect shorter wavelengths (UV light) [17]. These nanoparticles have been used in sunscreen preparations to protect human skin successfully. They are particularly effective in blocking the whole UVA spectrum, which may not be covered adequately by other chemical sunscreens [18]. However, TiO_2 and ZnO are large microscopic particles that appear as white pastes. The reduction of the size of such particles to the nanoscale renders them transparent while still absorbing UV radiation, resulting in significant aesthetic improvement [18]. A topically applied nano-Pts gel, formulated by mixing nano-Pts in water and carboxyvinyl polymer, was shown to protect mouse skin against UVB-induced inflammation [19].

Furthermore, nano-Pts had an anti-apoptotic effect against UVB irradiation in vitro as well as *in vivo*. It has been shown that the pretreatment of HaCaT keratinocytes with nano-Pts caused a significant inhibition of UVB- and UVC-induced apoptosis that is associated with a decrease in the UV-induced ROS production [19] (Figure 1). Therefore, nano-Pts can inhibit the inflammatory responses and apoptosis in skin. The nano-Pts gel is also useful with photoallergy. UVA-induced inflammation was clinically and histologically reduced by the application of nano-Pts gel. Proinflammatory cytokine levels, such as TNF-α, IL-1β and MIF in skin tissues were decreased by nano-Pts treatment in mice with photoallergy. Notably, nano-Pts may have interesting anti-skin aging properties.

In conclusion, nano-Pts seem to be beneficial in the protection against UV irradiation due to their ability to scavenge ROS and their aesthetic improvement of preparations.

4. Protective Effect of nano-Pts against RANKL-Stimulated Osteoclastogenesis and Cigarette Smoke-induced Pulmonary Inflammation

The receptor activator of NF-κB (RANK) ligand (RANKL) is one of the most essential cytokine required for the process of osteoclast differentiation and activation [20]. RANKL is a key member of the tumor necrosis factor (TNF) family and is expressed as a membrane-bound protein in osteoblasts and stromal cells [21]. The fundamental role of these molecules in the bone metabolism is well demonstrated by the high bone mass and osteoclast deficiency in mice lacking RANKL or its receptor RANK [22-24].

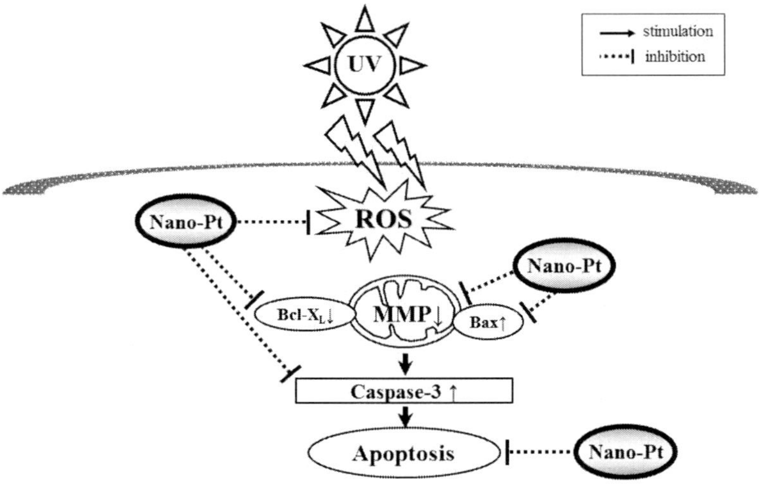

Figure 1. Schematic view of the suppressive effect of UV-induced ROS and apoptosis by nano-Pts in keratinocyte. UV increased ROS production, decreased MMP and resulted in caspase-3 activation. Nano-Pts pretreatment reversed all the observed endpoints. The solid lines indicate stimulation and the dotted lines indicate inhibition.

Clinical evidence also showed that the inhibition of RANKL activity significantly suppresses bone resorption both clinically and in mouse models. Nano-Pts significantly inhibit RANKL-stimulated osteoclastogenesis and LPS-induced inflammation in RAW 264.7 cells by the suppression of ROS. Treatment of RAW cells undergoing osteoclast differentiation induced by RANKL with Nano-Pts caused a significant decrease in the number of osteoclasts, in a concentration–dependent manner [25]. Moreover, increased oxidative stress has been considered as the important mechanism of the pulmonary inflammation in cigarette smokers. Cigarette smoke generates a huge amount of ROS, mainly superoxide radicals, hydrogen peroxide, hydroxyl radicals. All these are toxic to lungs and inflammation [26, 27]. Antioxidants may used as the effective therapeutic agents for inflammatory lung diseases induced by smoking such as chronic obstructive pulmonary disease [28, 29]. Nano-Pts are well known for their antioxidant activity, the intranasal administration of nano-Pts prior to cigarette smoke exposure has been shown to inhibit inflammation in the lungs of mice.

5. Protective Effect of nano-Pts against Hyperthermia Induced Inflammation

Hyperthermia (HT) is a form of heat shock treatment that can induce apoptosis in several cells, especially cancer cells. The induction of apoptosis has been proposed as a mechanism for heat-induced cell killing. ROS plays a very important role in apoptosis induction under both physiological and pathological conditions. Many studies have reported that oxidative stress plays a vital role in HT-induced apoptosis. Exposure to heat stress increases ROS (i.e. $O_2^{\cdot -}$, H_2O_2 and nitric oxide) in various cell lines and tumor tissues [30-32]. Furthermore, 44°C HT induces elevation of the intracellular Ca^{2+} concentration ($[Ca^{2+}]_i$) which is associated with enhanced HT-induced apoptosis [33, 34]. For example, in chondrocytes ROS causes a rise in $[Ca^{2+}]_i$ in cytoplasm, which induces calcium influx into mitochondria and nuclei to control apoptosis [35]. Also, in human myelomonocytic lymphoma U937 cells, HT-induced apoptosis is associated with intracellular generation of $O_2^{\cdot -}$, rapid lipid peroxidation, and an increase in intracellular $[Ca^{2+}]_i$ [36]. In the latter cell line, the pretreatment with nano-Pts suppressed HT-induced apoptosis in a concentration-dependent manner. Interestingly, only $O_2^{\cdot -}$, but not peroxides, were suppressed in presence of nano-Pts. This is in contrast to the reported durability of nano-Pts in scavenging ROS and thus this selective suppression of ROS by the pretreatment with nano-Pts was suggested to be a consequence of the reversal of HT-induced apoptosis rather than being a causative for the inhibition.

As for the underlying molecular mechanisms involved, it is well known that apoptosis is often associated with the loss of mitochondrial membrane potential (MMP) and the expression of Fas receptors, which are considered as endpoints for apoptosis. There was a restoration of the decreased MMP and a tendency for a reduction in Fas expression and content in the presence of nano-Pts in U937 cells. The activation of caspases, which are central executors of apoptosis, is associated with proteolytic activity that collectively contributes to the apoptotic program [37]. In addition, the proapoptotic Bcl-2 family members play a role in the regulation of the mitochondrial apoptotic pathway where their cleavage and subsequent translocation into the mitochondria promote the release of cytochrome *c* [38]. The expression of Bcl-2 family proteins, such as Bid, decreased after HT treatment but increased following pre-incubation with nano-Pts. Moreover, caspase-3 activation by HT treatment was also inhibited in the presence of nano-Pts [39].

Conclusion

ROS play a significant role in the inflammatory response. During an inflammatory response, the excessive production of ROS can cause major damage to cells, which can lead to DNA damage and mutations [40, 41]. Various factors involved in the inflammatory process are physical and chemical stimulants that are released during the immune response, and by tissue necrosis. Recently, we developed nano-Pts stabilized with polyacrylate to form a stable colloid solution. These nano-Pts are potent free radical scavengers and SOD/catalase mimetics [2]. Nanoparticles can be antioxidants, thus reducing the production of ROS in the body.

In conclusion, the benefits of nano Pts suggest the possibility of their use in sunscreen products for human skin protection, and progression to chemical trials in the near future by providing effective protection against inflammatory responses

Acknowledgments

These experiments were carried out in collaboration with researchers whose work appears in the References.

References

[1] Toshima N and Yonezawa T. Bimetallic nanoparticlesnovel materials for chemical and physical applications. *New J Chem* 1998;22: 1179-201.

[2] Kajita M, Hikosaka K, Iitsuka M, Kanayama A, Toshima N, Miyamoto Y. Platinum nanoparticle is a useful scavenger of superoxide anion and hydrogen peroxide. *Free Radic Res* 2007; 41: 615-26.

[3] Bhattacharya R and Mukherjee P. Biological properties of "naked" metal nanoparticles. *Adv Drug Deliv Rev* 2008; 60: 1289-306.

[4] Onizawa S, Aoshiba K, Kajita M, Miyamoto Y, Nagai A. Platinum nanoparticle antioxidants inhibit pulmonary inflammation in mice exposed to cigarette smoke. *Pulm Pharmacol Ther* 2009; 22: 340-9.

[5] Asharani PV, Xinyi N, Hande MP, Valiyaveettil S. DNA damage and p53-mediated growth arrest in human cells treated with platinum nanoparticles. *Nanomedicine (Lond)* 2010;5:51-64.

[6] Gao J, Liang G, Zhang B, Kuang Y, Zhang X, Xu B. FePt@CoS(2) yolk-shell nanocrystals as a potent agent to kill HeLa cells. *J Am Chem Soc* 2007; 129: 1428-33.

[7] Matsumura Y and Ananthaswamy HN. Short-term and long-term cellular and molecular events following UV irradiation of skin: implications for molecular medicine. *Expert Rev Mol Med* 2002; 4: 1-22. Review.

[8] Tenkate TD. Occupational exposure to ultraviolet radiation: a health risk assessment. *Rev Environ Health* 1999; 14: 187-209.

[9] Peus D, Vasa RA, Meves A, Pott M, Beyerle A, Squillace K *et al.* H_2O_2 is an important mediator of UVB-induced EGF-receptor phosphorylation in cultured keratinocytes. *J Invest Dermatol* 1998; 110: 966-71.

[10] Matsumura Y and Ananthaswamy HN. Toxic effects of ultraviolet radiation on the skin. *Toxicol Appl Pharmacol* 2004; 195: 298-308.

[11] Bokov A, Chaudhuri A, Richardson A. The role of oxidative damage and stress in aging. *Mech Ageing Dev* 2004; 125: 811-26.

[12] Halliwell B. Antioxidant defence mechanisms: from the beginning to the end (of the beginning). *Free Radic Res* 1999; 31: 261-72.

[13] Saija A, Tomaino A, Trombetta D, Giacchi M, De Pasquale A, Bonina F. Influence of different penetration enhancers on in vitro skin permeation and in vivo photoprotective effect of flavonoids. *Int J Pharm* 1998; 175: 85-94.

[14] Casagrande R, Georgetti SR, Verri WA Jr, Dorta DJ, dos Santos AC, Fonseca MJ. Protective effect of topical formulations containing quercetin against UVB-induced oxidative stress. *J Photochem Photobiol B* 2006: 84: 21-7.

[15] Myllylä R, Majamaa K, Günzler V, Hanauske-Abel HM, Kivirikko KI. Ascorbate is consumed stoichiometrically in the uncoupled reactions catalyzed by prolyl 4-hydroxylase and lysyl hydroxylase. *J Biol Chem* 1984; 259: 5403-5.

[16] Camera E, Mastrofrancesco A, Fabbri C, Daubrawa F, Picardo M, Sies H *et al.* Astaxanthin, canthaxanthin and beta-carotene differently affect UVA-induced oxidative damage and expression of oxidative stress-responsive enzymes. *Exp Dermatol* 2009; 18: 222-31.

[17] Sadrieh N, Wokovich AM, Gopee NV, Zheng J, Haines D, Parmiter D *et al.* Lack of significant dermal penetration of titanium dioxide from sunscreen formulations containing nano- and submicron-size TiO_2 particles. *Toxicol Sci* 2010; 115: 156-66.

[18] More BD. Physical sunscreens: on the comeback trail. *Indian J Dermatol Venereol Leprol* 2007; 73: 80-5.

[19] Yoshihisa Y, Honda A, Zhao QL, Makino T, Abe R, Matsui K *et al.* Protective effects of platinum nanoparticles against UV-light-induced epidermal inflammation. *Exp Dermatol* 2010; 19: 1000-6.

[20] Vaananen HK, Laitala-Leinonen T. Osteoclast lineage and function. *Arch Biochem Biophys.* 2008; 473: 132-8.

[21] Boyle WJ, Simonet WS, Lacey DL. Osteoclast differentiation and activation. *Nature* 2003; 423: 337-42.

[22] Teitelbaum SL. Bone resorption by osteoclasts. *Science* 2000; 289: 1504-8.

[23] Yang S, Ries WL, Key LL. Superoxide generation in transformed B-lymphocytes from patients with severe, malignant osteopetrosis. *Mol Cell Biochem* 1999; 199: 15-24.

[24] Suzuki A, Sekiguchi S, Asano S, Itoh M. Pharmacological topics of bone metabolism: recent advances in pharmacological management of osteoporosis. *J Pharmacol Sci* 2008; 106: 530-5.

[25] Nomura M, Yoshimura Y, Kikuiri T, Hasegawa T, Taniguchi Y, Deyama Y, Koshiro K, Sano H, Suzuki K, Inoue N. Platinum nanoparticles suppress osteoclastogenesis through scavenging of reactive oxygen species produced in RAW264.7 cells. *J Pharmacol Sci* 2011; 117: 243-52.

[26] Church DF, Pryor WA. Free-radical chemistry of cigarette smoke and its toxicological implications. *Environ Health Perspect* 1985; 64:111-26.

[27] Pryor WA, Stone K. Oxidants in cigarette smoke. Radicals, hydrogen peroxide, peroxynitrate, and peroxynitrite. *Ann N Y Acad Sci* 1993; 686: 12-7.

[28] Owen CA. Proteinases and oxidants as targets in the treatment of chronic obstructive pulmonary disease. *Proc Am Thorac Soc* 2005; 2:373-85.

[29] Yoshida T, Tuder RM. Pathobiology of cigarette smoke-induced chronic obstructive pulmonary disease. *Physiol Rev* 2007; 87: 1047-82.

[30] Yoshikawa T, Naito Y, Nakamura S, Nishimura S, Kaneko T, Iinuma S *et al.* Effect of rebamipide on lipid peroxidation and gastric mucosal injury induced by indometacin in rats. *Arzneimittelforschung* 1993; 43: 1327-30.

[31] Davidson JF, Whyte B, Bissinger PH, Schiestl RH. Oxidative stress is involved in heat-induced cell death in Saccharomyces cerevisiae. *Proc Natl Acad Sci U S A* 1996; 93: 5116-21.

[32] Flanagan SW, Moseley PL, Buettner GR. Increased flux of free radicals in cells subjected to hyperthermia: detection by electron paramagnetic resonance spin trapping. *FEBS Lett* 1998; 431: 285-6.

[33] Zhao QL, Fujiwara Y, Kondo T. Mechanism of cell death induction by nitroxide and hyperthermia. *Free Radic Biol Med* 2006; 40: 1131-43.

[34] Zhao QL, Kondo T, Noda A, Fujiwara Y. Mitochondrial and intracellular free-calcium regulation of radiation-induced apoptosis in human leukemic cells. *Int J Radiat Biol* 1999; 75: 493-504.

[35] Asada S, Fukuda K, Nishisaka F, Matsukawa M, Hamanisi C. Hydrogen peroxide induces apoptosis of chondrocytes; involvement of calcium ion and extracellular signal-regulated protein kinase. *Inflamm Res* 2001; 50: 19-23.

[36] Cui ZG, Kondo T, Matsumoto H. Enhancement of apoptosis by nitric oxide released from alpha-phenyl-tert-butyl nitrone under hyperthermic conditions. *J Cell Physiol* 2006; 206: 468-76.

[37] Chang HY and Yang X. Proteases for cell suicide: functions and regulation of caspases. *Microbiol Mol Biol Rev* 2000; 64: 821-46.

[38] Li H, Zhu H, Xu CJ, Yuan J. Cleavage of BID by caspase 8 mediates the mitochondrial damage in the Fas pathway of apoptosis. *Cell* 1998; 94: 491-501.

[39] Yoshihisa Y, Zhao QL, Hassan MA et al. A SOD/catalase mimetic platinum nanoparticle inhibits heat-induced apoptosis in human lymphoma U937 and HH cells. *Free Radical Res* 2011; 45: 326-35.

[40] Pallavi Lonkar and Peter C.Dedon. Reactive species and DNA damage in chronic inflammation. *Int J Cancer* 2011; 128: 1999-2009.

Zaidi SF, Ahmed K, Yamamoto T, Kondo T, Usmanghani K, Kadowaki M, Sugiyama T. Effect of resveratrol on Helicobacter pylori-induced interleukin-8 secretion, reactive oxygen species generation and morphological changes in human gastric epithelial cells. *Bio Pharm Bull* 2009; 32: 1931-5.

In: Platinum: Compounds, Production and Applications
Editors: L.Varennikov and E. Yedemsky
ISBN: 978-1-62257-939-6

Chapter 6

PLATINUM-HALIDE SQUARE-PRISM COMPOUNDS: NOVEL GROUND STATES AND THEIR OPTICAL FEATURES

Shoji Yamamoto*and Jun Ohara
Department of Physics, Hokkaido University,
Sapporo, Japan

Abstract

Quasi-one-dimensional platinum-halide compounds are known by their unique optical properties and great effort is still devoted to fabricating their new varieties. Motivated by the newly designed square-prism-shaped iodo platinum complexes of general formula $[Pt(C_2H_8N_2)(C_{10}H_8N_2)I]_4(NO_3)_8 \cdot nH_2O$, we construct a microscopic theory of their electronic structure with particular emphasis on the symmetry properties and in consideration of the experimental observations. We reveal a variety of charge-density-wave states and illuminate their optical features.

PACS 71.45.Lr, 71.10.Hf, 78.20.Bh, 78.20.Ci

Keywords: Platinum-halide square-prism compound, Charge density wave, Optical conductivity

1. Introduction

Quasi-one-dimensional halogen (X)-bridged transition-metal (M) complexes [1, 2, 3, 4, 5, 6] have been fascinating numerous chemists and physicists for more than a century. Their performance consists of competing electron-electron and electron-lattice interactions [7, 8, 9], being spiced with low dimensionality and d-p hybridization [10, 11]. The earliest example $[Pt(ea)_4Cl]Cl_2 \cdot 2H_2O$ (ea = ethylamine = C_2H_7N) [12] and its bromo derivative [13], which are well-known as Wolfram's red and Reihlen's green salts, respectively, exhibit a Peierls-distorted mixed-valent ground state [14]. They are full of intriguing optoelectronic properties [15] such as intense and dichroic charge-transfer absorption, strong

*E-mail address: yamamoto@phys.sci.hokudai.ac.jp

resonance enhancement of Raman spectra, and luminescence with large Stokes shift. While palladium analogs stay mixed-valent [16], nickel analogs instead take a Mott-insulating monovalent regular chain structure [17, 18]. Microscopic electronic-structure calculations demonstrated the robustness [19] and tunability [20] of the Peierls instability. Intrinsic defect states such as solitons and polarons were also calculated [10, 11, 21, 22, 23] and indeed observed [24, 25, 26, 27].

Metal binucleation leads to a wider variety of electronic states [28, 29, 30, 31]. Diplatinum-halide chain compounds, $R_4[Pt_2(pop)_4X]\cdot nH_2O$ [X = Br, I; pop = diphosphonate = $P_2O_5H_2$; R = K, $(C_2H_5)_2NH_2$] [32, 33], have a ground state with halogen-sublattice dimerization, which is reminiscent of the MX conventional, while their analog without any counter ion, $Pt_2(dta)_4I$ (dta = dithioacetate = CH_3CS_2) [34], possesses a novel ground state with metal-sublattice dimerization [35], which may be due in part to the dta ligands being twisted [36]. The latter compound exhibits another mixed-valent state without cell doubling at higher temperatures [35, 37], whereas the former compound is less varied thermally but tunable by light irradiation and/or pressure application [38, 39, 40, 41, 42]. Local excitations on such backgrounds [43, 44] are all the more interesting because the metal-metal direct bonding significantly improves the transport properties [45]. $Pt_2(dta)_4I$, or $Pt_2(RCS_2)_4I$ (R = alkyl chain = C_nH_{2n+1}) [46] in more general, indeed exhibits metallic conduction above room temperature [35, 46]. Neutral and charged solitons may simultaneously be stabilized and survive longer lifetime in MMX chains [47, 48]. Recent electron-spin-resonance experiments have actually detected mobile spins being thermally excited [49] and/or photogenerated [50] in iodo-bridged dithioalkyl-ligand diplatinum complexes, which are reasonably attributable to neutral solitons. Nickel analogs [51] are available in this class as well and they bring about distinct observations [52, 53, 54, 55]. There are further attempts [56, 57] to bridge polynuclear or heterometallic units by halogens.

Hundreds of MX compounds were thus synthesized and studied, but they all have single-chain-assembled structures. In such circumstances, coordination chemists took a step toward bringing into interaction two or more MX chains. Initial examples, $(bpy)[Pt(dien)Br]_2Br_4 \cdot 2H_2O$ (dien = diethylentriamine = $C_4H_{13}N_3$; bpy = 4, 4′-bipyridyl = $C_{10}H_8N_2$) [58], and $(\mu\text{-bpym})[Pt(en)X]_2X(ClO_4)_3\cdot H_2O$ (X = Cl, Br; en = ethylendiamine = $C_2H_8N_2$; μ-bpym = 2, 2′-bipyrimidine = $C_8H_6N_4$) [59], both crystallize in a ladder structure, where binuclear units, $\{(bpy)[Pt(dien)]_2\}$ and $\{(\mu\text{-bpym})[Pt(en)]_2\}$ are, respectively, stacked with bridging halogen atoms in between. They are made in distinct ground states of mixed valence [60]. The former comprises a couple of charge-density-waved (CDWd) chains out of phase (OP), while the latter consists of those in phase (IP) [61, 62]. They further form a striking contrast against photoirradiation. Photoinduced intrachain charge-transfer excitons are self-stabilized in both materials. However, they are balanced on double chains in the case of an IP-CDWd background, whereas they are localized onto one leg in the case of an OP-CDWd background [63, 64]. When the fully-trapped excitons decay by luminescence, the Stokes shift amounts to half the optical gap in the OP-CDW case, while it is no more than quarter the optical gap in the IP-CDW case. The former finding is almost equal to those in isolated MX chains [65].

Further chemical explorations of new MX varieties have led to a novel example of a metal-organic nanotubular assembly, $[Pt(en)(bpy)I]_4(NO_3)_8 \cdot 16H_2O$ [66], where tetranuclear frames, [Pt(en)(bpy)]$_4$, together with bridging iodine atoms, are polymerized into a square-prism structure. Its structural components are exchangeable because it is a bottom-up realization. The organic ligands are decisive of the inside diameter and thus influential for the porosity. The substitution of chlorine for iodine is likely to enhance the Peierls distortion [67], whereas that of nickel for platinum may realize a Mott-insulating spin tube [68]. Metal-halide tubes must be a fascinating stage featuring well-correlated electrons coupled with phonons on a way from one to two dimensions. A voyage of exploration into this brand-new MX family is fairly under way. We present a hot report on its ground-state properties.

2. Group-Theoretical Analysis

We establish a two-band extended Peierls-Hubbard Hamiltonian for describing metal-halide square-prism compounds,

$$
\begin{aligned}
\mathcal{H} = & \sum_{l=1}^{4}\sum_{n=1}^{N}\sum_{s=\uparrow,\downarrow}\Big\{\big[t_{MX}^{\parallel}-\alpha(u_{n+1:lM}^{\parallel}-u_{n:lX}^{\parallel})\big]a_{n+1:lMs}^{\dagger}a_{n:lXs} \\
& -\big[t_{MX}^{\parallel}-\alpha(u_{n:lX}^{\parallel}-u_{n:lM}^{\parallel})\big]a_{n:lXs}^{\dagger}a_{n:lMs}-t_{MM}^{\perp}a_{n:l+1Ms}^{\dagger}a_{n:lMs}+\mathrm{H.c.}\Big\} \\
& +\sum_{l=1}^{4}\sum_{n=1}^{N}\sum_{s=\uparrow,\downarrow}\Big\{\big[\varepsilon_M-\beta_M(u_{n:lX}^{\parallel}-u_{n-1:lX}^{\parallel})\big]n_{n:lMs} \\
& +\big[\varepsilon_X-\beta_X(u_{n+1:lM}^{\parallel}-u_{n:lM}^{\parallel})\big]n_{n:lXs}\Big\} \\
& +\sum_{l=1}^{4}\sum_{n=1}^{N}\sum_{s,s'=\uparrow,\downarrow}\Big\{\sum_{A=M,X}\Big[\frac{U_A}{4}n_{n:lAs}n_{n:lA-s}+V_{AA}^{\parallel}n_{n:lAs}n_{n+1:lAs'} \\
& +V_{AA}^{\perp}n_{n:lAs}n_{n:l+1As'}+\frac{V_{AA}^{\times}}{2}n_{n:lAs}n_{n:l+2As'} \\
& +V_{AA}^{\times\!\times}(n_{n:lAs}n_{n+1:l+1As'}+n_{n:l+1As}n_{n+1:lAs'})\Big] \\
& +V_{MX}^{\parallel}(n_{n:lMs}n_{n:lXs'}+n_{n:lXs}n_{n+1:lMs'})+V_{MX}^{\times\!\times}(n_{n:lMs}n_{n:l+1Xs'} \\
& +n_{n:l+1Ms}n_{n:lXs'}+n_{n:lXs}n_{n+1:l+1Ms'}+n_{n:l+1Xs}n_{n+1:lMs'})\Big\} \\
& +\sum_{l=1}^{4}\sum_{n=1}^{N}\frac{K_{MX}^{\parallel}}{2}\big[(u_{n:lX}^{\parallel}-u_{n:lM}^{\parallel})^2+(u_{n+1:lM}^{\parallel}-u_{n:lX}^{\parallel})^2\big], \qquad (1)
\end{aligned}
$$

where n, l, A, and s specify the M_4X_4 unit of rectangular parallelepiped, MX chain leg, site, and spin, respectively. All the model parameters are detailed in Fig. 1. Defining the Fourier transformation as

$$a_{k:lAs}=\frac{1}{\sqrt{N}}\sum_{n}e^{-i\boldsymbol{k}\cdot\boldsymbol{c}(n+\delta_{A,X}/2)}a_{n:lAs}, \qquad (2)$$

$$u_{k:lA}^{\parallel}=\frac{1}{\sqrt{N}}\sum_{n}e^{-i\boldsymbol{k}\cdot\boldsymbol{c}(n+\delta_{A,X}/2)}u_{n:lA}^{\parallel}, \qquad (3)$$

we rewrite the Hamiltonian compactly in the momentum space as

$$
\begin{aligned}
\mathcal{H} = & K_{MX}^{\|}\sum_{A}\sum_{l}\sum_{Q}\Big\{|u_{Q:lA}^{\|}|^2 - \frac{1}{2}(u_{Q:lM}^{\|}u_{Q:lX}^{\|*} + \mathrm{c.c.})\cos\frac{Qc}{2}\Big\} \\
& + \sum_{A,A'}\sum_{l,l'}\sum_{k,Q}\sum_{s,s'}\langle k+Q:lA,s|t|k:l'A',s'\rangle a_{k+Q:lAs}^{\dagger}a_{k:l'A's'} + \frac{1}{2}\sum_{A,A',B,B'} \\
& \times \sum_{l,l',m,m'}\sum_{k,k',Q}\sum_{s,s',t,t'}\langle k+Q:lA,s;k':mB,t|v|k:l'A',s';k'+Q:m'B',t'\rangle \\
& \times a_{k+Q:lAs}^{\dagger}a_{k':mBt}^{\dagger}a_{k'+Q:m'B't'}a_{k:l'A's'},
\end{aligned} \tag{4}
$$

where the spin indices in the momentum-dependent coefficients of one and two bodies can be omitted as

$$
\langle k+Q:lA,s|t|k:l'A',s'\rangle = \langle k+Q:lA|t|k:l'A'\rangle\delta_{s,s'}, \tag{5}
$$

$$
\begin{aligned}
&\langle k+Q:lA,s;k':mB,t|v|k:l'A',s';k'+Q:m'B',t'\rangle \\
&= \langle k+Q:lA;k':mB|v|k:l'A';k'+Q:m'B'\rangle\delta_{s,s'}\delta_{t,t'}.
\end{aligned} \tag{6}
$$

Their explicit forms are readily available through the Fourier transform of the original Hamiltonian (1) and given in Appendix A. We decompose Eq. (4) within the Hartree-Fock (HF) scheme as

$$
\begin{aligned}
\mathcal{H}_{\mathrm{HF}} &= \sum_{l,l'}\sum_{A,A'}\sum_{k,Q}\sum_{s,s'}\sum_{\lambda=0,x,y,z} x_{lA,l'A'}^{\lambda}(Q;k)a_{k+Q:lAs}^{\dagger}a_{k:l'A's'}\sigma_{ss'}^{\lambda} \\
&\equiv \sum_{Q}\sum_{\lambda} h_Q^{\lambda},
\end{aligned} \tag{7}
$$

where the nonmagnetic ($\lambda = 0$) and magnetic ($\lambda = x, y, z$) variational parameters read as

$$
\begin{aligned}
x_{lA,l'A'}^{0}(Q;k) = & \langle k+Q:lA|t|k:l'A'\rangle \\
& + \frac{1}{2}\sum_{m,m'}\sum_{k'}\sum_{B,B'} e^{iQc\delta_{B,X}}\rho_{mB,m'B'}^{0}(Q;k') \\
& \times\big[2\langle k+Q:lA;k':mB|v|k:l'A';k'+Q:m'B'\rangle \\
& +2\langle k':mB;k+Q:lA|v|k'+Q:m'B';k:l'A'\rangle \\
& -\langle k':mB;k+Q:lA|v|k:l'A';k'+Q:m'B'\rangle \\
& -\langle k+Q:lA;k':mB|v|k'+Q:m'B';k:l'A'\rangle\big],
\end{aligned} \tag{8}
$$

$$
\begin{aligned}
x_{lA,l'A'}^{\lambda}(Q:k) = & -\frac{1}{2}\sum_{m,m'}\sum_{k'}\sum_{B,B'} e^{iQc\delta_{B,X}}\rho_{mB,m'B'}^{\lambda}(Q;k') \\
& \times\big[\langle k':mB;k+Q:lA|v|k:l'A';k'+Q:m'B'\rangle \\
& +\langle k+Q:lA;k':mB|v|k'+Q:m'B';k:l'A'\rangle\big],
\end{aligned} \tag{9}
$$

with the density matrices

$$
\rho_{lA,l'A'}^{\lambda}(Q;k) = \frac{1}{2}\sum_{s,s'}\langle c_{k+Q:lA,s}^{\dagger}c_{k:l'A',s'}\rangle_{\mathrm{HF}}\sigma_{s,s'}^{\lambda}. \tag{10}
$$

Here $\langle\ \rangle_{\rm HF}$ means taking the quantum average in a HF eigenstate with $u^{\parallel}_{Q:lA}$ minimizing the HF energy.

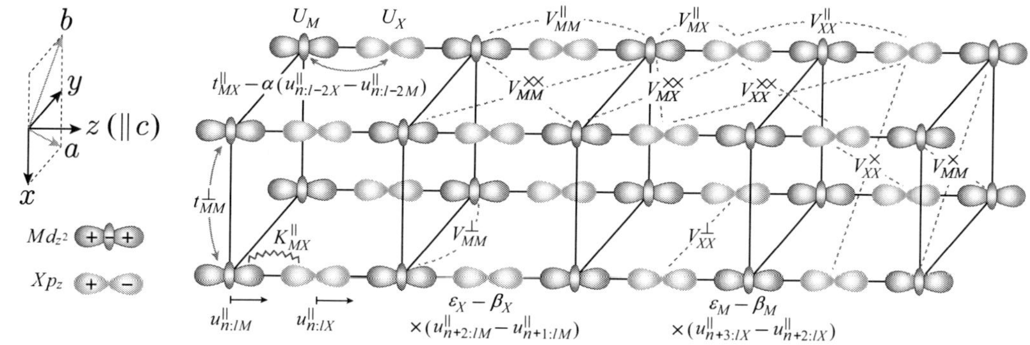

Figure 1. (Color online) Modelling of an MX square prism, where heavily and lightly shaded clouds denote $M\ d_{z^2}$ and $X\ p_z$ orbitals, the electron numbers on which are given by $n_{n:lMs} \equiv a^{\dagger}_{n:lMs}a_{n:lMs}$ and $n_{n:lXs} \equiv a^{\dagger}_{n:lXs}a_{n:lXs}$, respectively. The on-site energies of isolated atoms are given by ε_M and ε_X, while the electron hoppings between these levels are modelled by $t^{\parallel}_{MX}$ and $t^{\perp}_{MM}$. The on-site Coulomb interactions are refereed to as U_A $(A = M, X)$, whereas the interchain and intrachain different-site Coulomb interactions are specified with $V^{\perp}_{AA'}$, $V^{\times\times}_{AA'}$, $V^{\times}_{AA'}$, and $V^{\parallel}_{AA'}$ $(A, A' = M, X)$. The leg-direction displacements of metal and halogen ions, $u^{\parallel}_{n:lM}$ and $u^{\parallel}_{n:lX}$, interact with electrons through intersite (α) and intrasite (β_M, β_X) coupling constants at the cost of elastic energy in proportion to $K^{\parallel}_{MX}$.

Unless the gauge symmetry is broken, the symmetry group of any lattice electron system can be written as [69, 70]

$$\mathbf{G} = \mathbf{P} \times \mathbf{S} \times \mathbf{T}, \tag{11}$$

where $\mathbf{P}$, $\mathbf{S}$, and $\mathbf{T}$ are the groups of space, spin rotation, and time reversal, respectively. The space group is further decomposed as

$$\mathbf{P} = \mathbf{L} \wedge \mathbf{D}, \tag{12}$$

where $\mathbf{L}$ and $\mathbf{D}$ are the translation and point groups, respectively. When we assemble MX complexes into a square-prism lattice as depicted in Fig. 1, $\mathbf{L} = \mathbf{L}_1$, where $\mathbf{L}_m = \{mlc;\, m \in \mathbf{N}, l \in \mathbf{Z}\}$, and $\mathbf{D} = \mathbf{D}_{4h} = \mathbf{D}_4 + I\mathbf{D}_4$, where $\mathbf{D}_4 = \{E, C_{4z}, C_{2z}, C_{4z}^{-1}, C_{2x}, C_{2y}, C_{2a}, C_{2b}\}$. An irreducible representation of $\mathbf{G}$ over the real number field, which is referred to as $\check{G}$, consists of those of $\mathbf{P}$, $\mathbf{S}$, and $\mathbf{T}$: $\check{G} = \check{P} \otimes \check{S} \otimes \check{T}$. Once a wave vector $\boldsymbol{Q} \equiv (0, 0, Q)$ is fixed, the little group $\mathbf{D}(\boldsymbol{Q})$, for which $\boldsymbol{Q}$ is invariant, is correspondingly given. $\check{P}$ is therefore labeled $\boldsymbol{Q}\check{D}(\boldsymbol{Q})$. Since we take much interest in cell-doubled phases, we consider $Q = \pi/c$ as well as $Q = 0$, which are referred to as X and Γ, respectively, and whose little groups are both $\mathbf{D}_{4h}$ itself. The spin rotation by θ around an arbitrary axis $\boldsymbol{e}$ reads as $\sigma^0\cos(\theta/2) - i\boldsymbol{\sigma}\cdot\boldsymbol{e}\sin(\theta/2) \equiv u(\boldsymbol{e}, \theta)$, where σ^0 is the 2×2 unit matrix, while $\boldsymbol{\sigma} \equiv (\sigma^x, \sigma^y, \sigma^z)$ is a vector composed of the Pauli matrices. So assigning a 3×3 orthogonal matrix to $u(\boldsymbol{e}, \theta)$ as to satisfy $u(\boldsymbol{e}, \theta)\sigma^{\lambda}u^{\dagger}(\boldsymbol{e}, \theta) = \sum_{\mu=x,y,z}[O(u(\boldsymbol{e}, \theta))]_{\lambda\mu}\sigma^{\mu}$ $(\lambda = x, y, z)$, we can express the relevant representations of $\mathbf{S}$ as $\check{S}^0(u(\boldsymbol{e},\theta)) = 1$ (singlet) and $\check{S}^1(u(\boldsymbol{e},\theta)) = O(u(\boldsymbol{e},\theta))$ (triplet). $\mathbf{T}$ is represented by $\check{T}^0(t) = 1$ (symmetric) and $\check{T}^1(t) = -1$ (antisymmetric). Considering that the platinum $5d$ electrons are moderately correlated, we restrict our discussion to non-magnetic solutions derived from $\Gamma\check{D}(\Gamma) \otimes \check{S}^0 \otimes \check{T}^0$ and $\mathrm{X}\check{D}(\mathrm{X}) \otimes \check{S}^0 \otimes \check{T}^0$, whose space groups turn out to be $\mathbf{L}_1 \wedge \mathbf{D}_{4\mathrm{h}}$ and $\mathbf{L}_2 \wedge \mathbf{D}_{4\mathrm{h}}$, respectively.

Once we comprehend group actions, we can project the Hamiltonian (7) onto each irreducible representation. $\boldsymbol{c} \in \mathbf{L}_1$ acts on the electron operators as $\boldsymbol{c} \cdot a^\dagger_{k:lAs} = e^{-ikc} a^\dagger_{k:lAs}$. Actions of $p \in \mathbf{D}_{4\mathrm{h}}$ can be expressed as $p \cdot a^\dagger_{k:lMs} = [A_{1g}(p)]_{11} a^\dagger_{pk:lMs}$ and $p \cdot a^\dagger_{k:lXs} = [A_{2u}(p)]_{11} a^\dagger_{pk:lXs}$, where $[\check{D}(p)]_{ij}$ is the (i,j)-element of the $\check{D}$ representation matrix for p. Those of $u(\boldsymbol{e},\theta) \in \mathbf{S}$ read as $u(\boldsymbol{e},\theta) \cdot a^\dagger_{k:lAs} = \sum_{s'} [u(\boldsymbol{e},\theta)]_{s's} a^\dagger_{k:lAs'}$, Those of $t \in \mathbf{T}$ are defined as $t \cdot a^\dagger_{k:lAs} = (-1)^{\delta_{s,\uparrow}} a^\dagger_{-k:lA-s}$. All the fixed-point subspaces can be calculated as

$$h^{\lambda\tau}_{Q\check{D}[i,j]} = \frac{d_{\check{D}}}{2g} \sum_{t\in\mathbf{T}} \check{T}^\tau(t) \sum_{p\in\mathbf{D}_{4\mathrm{h}}} [\check{D}(p)]^*_{ij} tp \cdot h^\lambda_Q, \tag{13}$$

where $g\,(=16)$ is the order of $\mathbf{D}_{4\mathrm{h}}$ and $d_{\check{D}}\,(\leq 2)$ is the dimension of its arbitrary irreducible representation $\check{D}$. The Hamiltonian (7) reads as

$$\mathcal{H}_{\mathrm{HF}} = \sum_{Q=\Gamma,\mathrm{X}} \sum_{\check{D}(Q)} \sum_{\lambda=0,x,y,z} \sum_{\tau=0,1} h^{\lambda\tau}_{Q\check{D}(Q)}. \tag{14}$$

Multidimensional representations are generally available in a variety of isotropy subgroups. Here, for the two-dimensional representations E_g and E_u, we consider only the *axial* isotropy subgroups discarding those of two-dimensional fixed point subspace. Every irreducible representation is guaranteed to yield a stable solution only when its isotropy subgroup possesses a one-dimensional fixed point subspace [71].

The thus-obtained irreducible representations are detailed in Tables 1 and 2 and visualized in Fig. 2. Every broken-symmetry Hamiltonian has a unique solution and gives density matrices of its own symmetry. The electron densities $d_{n:lA} \equiv \sum_s \langle a^\dagger_{n:lAs} a_{n:lAs}\rangle_{\mathrm{HF}}$, the spin densities $s^\lambda_{n:lA} \equiv \sum_{s,s'} \langle a^\dagger_{n:lAs} a_{n:lAs'}\rangle_{\mathrm{HF}} \sigma^\lambda_{ss'}/2$, the bond orders $p_{n:lA;n':l'A'} \equiv \sum_s \langle a^\dagger_{n:lAs} a_{n':l'A's}\rangle_{\mathrm{HF}}$, the spin bond orders $t^\lambda_{n:lA;n':l'A'} \equiv \sum_{s,s'} \langle a^\dagger_{n:lAs} a_{n':l'A's'}\rangle_{\mathrm{HF}} \sigma^\lambda_{ss'}/2$, and the optimum lattice displacements $u^{\parallel}_{n:lA}$ are all described in terms of the density matrices of definite symmetry via Eqs. (2) and (3). Thus and thus, we obtain Fig. 2.

While cell-doubled X solutions, including charge density waves (CDWs) in the metal sublattice, are of main interest at low temperatures, some of Γ solutions are also interesting in themselves. The mixed-valent states available in the ΓB_{2g} and ΓEu representations are all characterized by interchain charge polarization. Their constituent MX chains are not necessarily neutral even at three-quarter filling. Some are electron-rich, while some are electron-poor. □-M-X-CDW and x-CP are actually stabilized with increasing $V^\perp_{MM}$ and $V^\times_{MM}$, respectively, neither of which is, however, practical. A breakdown of the charge neutrality in each constituent chain is more likely in a triangular structure [68], which is of purely geometric origin.

X solutions are classified into three groups: charge density waves in the metal sublattice with the halogen sublattice being distorted, those in the halogen sublattice with the metal sublattice being distorted, and bond order waves without any charge oscillation, which are

Table 1. Symmetry properties of the irreducible representations $\Gamma\check{D}(\Gamma)\otimes\check{S}^0\otimes\check{T}^0$ available in the MX square-prism Hamiltonian (1) subject to axial isotropy subgroup. The first term of every broken-symmetry Hamiltonian gives each individual fixed-point subspace. Refer to Appendix B as well for further details of the relevant point groups

Irreducible representation	Axial isotropy subgroup	Broken-symmetry Hamiltonian	Descriptive abbreviation
$\Gamma A_{1g}\otimes\check{S}^0\otimes\check{T}^0$	$\mathbf{D}_{4h}\mathbf{L}_1\mathbf{ST}$	$h^{00}_{\Gamma A_{1g}[1,1]}$	PM
$\Gamma A_{2g}\otimes\check{S}^0\otimes\check{T}^0$	$\mathbf{C}_{4h}\mathbf{L}_1\mathbf{ST}$	$h^{00}_{\Gamma A_{2g}[1,1]}+h^{00}_{\Gamma A_{1g}[1,1]}$	∧-MX-BOW
$\Gamma B_{1g}\otimes\check{S}^0\otimes\check{T}^0$	$\mathbf{D}_{2h}\mathbf{L}_1\mathbf{ST}$	$h^{00}_{\Gamma B_{1g}[1,1]}+h^{00}_{\Gamma A_{1g}[1,1]}$	□-MM-XX-BOW
$\Gamma B_{2g}\otimes\check{S}^0\otimes\check{T}^0$	$\mathbf{D}_{2ah}\mathbf{L}_1\mathbf{ST}$	$h^{00}_{\Gamma B_{2g}[1,1]}+h^{00}_{\Gamma A_{1g}[1,1]}$	□-M-X-CDW
$\Gamma E_g(1)\otimes\check{S}^0\otimes\check{T}^0$	$\mathbf{C}_{2xh}\mathbf{L}_1\mathbf{ST}$	$h^{00}_{\Gamma E_g[1,1]}$ $+h^{00}_{\Gamma B_{1g}[1,1]}+h^{00}_{\Gamma A_{1g}[1,1]}$	$(+--+)$-MX-BOW
$\Gamma E_g(2)\otimes\check{S}^0\otimes\check{T}^0$	$\mathbf{C}_{2ah}\mathbf{L}_1\mathbf{ST}$	$\sum_{i,j}h^{00}_{\Gamma E_g[i,j]}/2$ $+h^{00}_{\Gamma B_{2g}[1,1]}+h^{00}_{\Gamma A_{1g}[1,1]}$	$(+0-0)$-MX-BOW
$\Gamma A_{1u}\otimes\check{S}^0\otimes\check{T}^0$	$\mathbf{D}_{4}\mathbf{L}_1\mathbf{ST}$	$h^{00}_{\Gamma A_{1u}[1,1]}+h^{00}_{\Gamma A_{1g}[1,1]}$	//-MX-BOW
$\Gamma A_{2u}\otimes\check{S}^0\otimes\check{T}^0$	$\mathbf{C}_{4v}\mathbf{L}_1\mathbf{ST}$	$h^{00}_{\Gamma A_{2u}[1,1]}+h^{00}_{\Gamma A_{1g}[1,1]}$	$(----)$-MX-BOW
$\Gamma B_{1u}\otimes\check{S}^0\otimes\check{T}^0$	$\mathbf{D}_{2d}\mathbf{L}_1\mathbf{ST}$	$h^{00}_{\Gamma B_{1u}[1,1]}+h^{00}_{\Gamma A_{1g}[1,1]}$	$(+-+-)$-MX-BOW
$\Gamma B_{2u}\otimes\check{S}^0\otimes\check{T}^0$	$\mathbf{D}_{2ad}\mathbf{L}_1\mathbf{ST}$	$h^{00}_{\Gamma B_{2u}[1,1]}+h^{00}_{\Gamma A_{1g}[1,1]}$	×-MX-BOW
$\Gamma E_u(1)\otimes\check{S}^0\otimes\check{T}^0$	$\mathbf{C}_{2xv}\mathbf{L}_1\mathbf{ST}$	$h^{00}_{\Gamma E_u[1,1]}$ $+h^{00}_{\Gamma B_{1g}[1,1]}+h^{00}_{\Gamma A_{1g}[1,1]}$	x-CP
$\Gamma E_u(2)\otimes\check{S}^0\otimes\check{T}^0$	$\mathbf{C}_{2av}\mathbf{L}_1\mathbf{ST}$	$\sum_{i,j}h^{00}_{\Gamma E_u[i,j]}/2$ $+h^{00}_{\Gamma B_{2g}[1,1]}+h^{00}_{\Gamma A_{1g}[1,1]}$	a-CP

Table 2. Symmetry properties of the irreducible representations $\mathrm{X}\check{D}(\mathrm{X}) \otimes \check{S}^0 \otimes \check{T}^0$ available in the MX square-prism Hamiltonian (1) subject to axial isotropy subgroup. The first term of every broken-symmetry Hamiltonian gives each individual fixed-point subspace. Refer to Appendix B as well for further details of the relevant point groups

Irreducible representation	Axial isotropy subgroup	Broken-symmetry Hamiltonian	Descriptive abbreviation
$\mathrm{X}A_{1g} \otimes \check{S}^0 \otimes \check{T}^0$	$\mathbf{D}_{4h}\mathbf{L}_2\mathbf{ST}$	$h^{00}_{\mathrm{X}A_{1g}[1,1]} + h^{00}_{\Gamma A_{1g}[1,1]}$	$(++++)$-M-CDW
$\mathrm{X}A_{2g} \otimes \check{S}^0 \otimes \check{T}^0$	$(1 + C_{2x}\boldsymbol{c})$ $\times \mathbf{C}_{4h}\mathbf{L}_2\mathbf{ST}$	$h^{00}_{\mathrm{X}A_{2g}[1,1]} + h^{00}_{\Gamma A_{1g}[1,1]}$	$\wedge$-MM-BOW
$\mathrm{X}B_{1g} \otimes \check{S}^0 \otimes \check{T}^0$	$(1 + C_{2a}\boldsymbol{c})$ $\times \mathbf{D}_{2h}\mathbf{L}_2\mathbf{ST}$	$h^{00}_{\mathrm{X}B_{1g}[1,1]} + h^{00}_{\Gamma A_{1g}[1,1]}$	$\square$-MM-BOW
$\mathrm{X}B_{2g} \otimes \check{S}^0 \otimes \check{T}^0$	$(1 + C_{2x}\boldsymbol{c})$ $\times \mathbf{D}_{2ah}\mathbf{L}_2\mathbf{ST}$	$h^{00}_{\mathrm{X}B_{2g}[1,1]} + h^{00}_{\Gamma A_{1g}[1,1]}$	$(+-+-)$-M-CDW
$\mathrm{X}E_g(1) \otimes \check{S}^0 \otimes \check{T}^0$	$(1 + C_{2z}\boldsymbol{c})$ $\times \mathbf{C}_{2xh}\mathbf{L}_2\mathbf{ST}$	$h^{00}_{\mathrm{X}E_g[1,1]}$ $+h^{00}_{\Gamma B_{1g}[1,1]} + h^{00}_{\Gamma A_{1g}[1,1]}$	$(+--+)$-X-CDW
$\mathrm{X}E_g(2) \otimes \check{S}^0 \otimes \check{T}^0$	$(1 + C_{2z}\boldsymbol{c})$ $\times \mathbf{C}_{2ah}\mathbf{L}_2\mathbf{ST}$	$\sum_{i,j} h^{00}_{\mathrm{X}E_g[i,j]}/2$ $+h^{00}_{\Gamma B_{2g}[1,1]} + h^{00}_{\Gamma A_{1g}[1,1]}$	$(+0-0)$-X-CDW
$\mathrm{X}A_{1u} \otimes \check{S}^0 \otimes \check{T}^0$	$(1 + I\boldsymbol{c})$ $\times \mathbf{D}_4\mathbf{L}_2\mathbf{ST}$	$h^{00}_{\mathrm{X}A_{1u}[1,1]} + h^{00}_{\Gamma A_{1g}[1,1]}$	$\wedge$-XX-BOW
$\mathrm{X}A_{2u} \otimes \check{S}^0 \otimes \check{T}^0$	$(1 + I\boldsymbol{c})$ $\times \mathbf{C}_{4v}\mathbf{L}_2\mathbf{ST}$	$h^{00}_{\mathrm{X}A_{2u}[1,1]} + h^{00}_{\Gamma A_{1g}[1,1]}$	$(++++)$-X-CDW
$\mathrm{X}B_{1u} \otimes \check{S}^0 \otimes \check{T}^0$	$(1 + I\boldsymbol{c})$ $\times \mathbf{D}_{2d}\mathbf{L}_2\mathbf{ST}$	$h^{00}_{\mathrm{X}B_{1u}[1,1]} + h^{00}_{\Gamma A_{1g}[1,1]}$	$(+-+-)$-X-CDW
$\mathrm{X}B_{2u} \otimes \check{S}^0 \otimes \check{T}^0$	$(1 + I\boldsymbol{c})$ $\times \mathbf{D}_{2ad}\mathbf{L}_2\mathbf{ST}$	$h^{00}_{\mathrm{X}B_{2u}[1,1]} + h^{00}_{\Gamma A_{1g}[1,1]}$	$\square$-XX-BOW
$\mathrm{X}E_u(1) \otimes \check{S}^0 \otimes \check{T}^0$	$(1 + I\boldsymbol{c})$ $\times \mathbf{C}_{2xv}\mathbf{L}_2\mathbf{ST}$	$h^{00}_{\mathrm{X}E_u[1,1]}$ $+h^{00}_{\Gamma B_{1g}[1,1]} + h^{00}_{\Gamma A_{1g}[1,1]}$	$(++--)$-M-CDW
$\mathrm{X}E_u(2) \otimes \check{S}^0 \otimes \check{T}^0$	$(1 + I\boldsymbol{c})$ $\times \mathbf{C}_{2av}\mathbf{L}_2\mathbf{ST}$	$\sum_{i,j} h^{00}_{\mathrm{X}E_u[i,j]}/2$ $+h^{00}_{\Gamma B_{2g}[1,1]} + h^{00}_{\Gamma A_{1g}[1,1]}$	$(0+0-)$-M-CDW

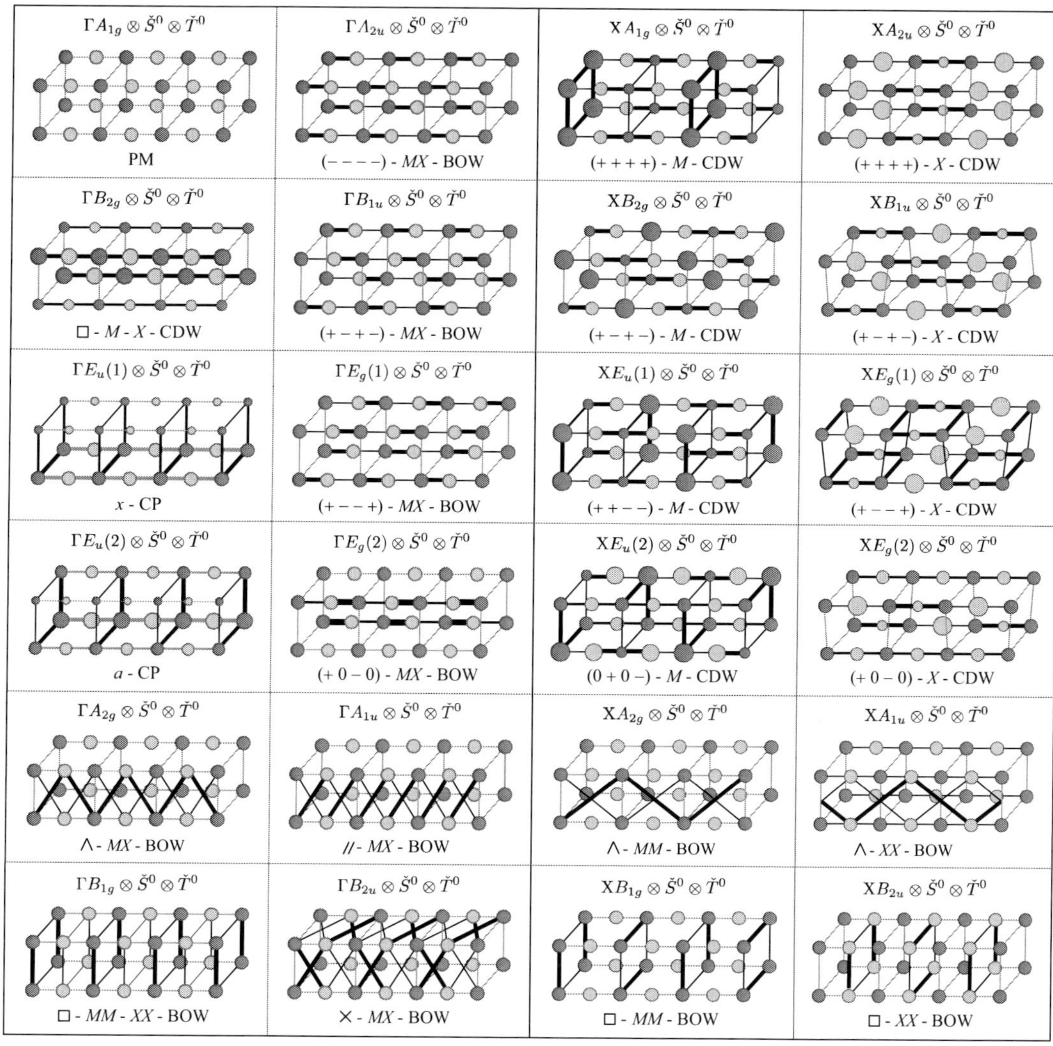

Figure 2. (Color online) Possible electronic states of the $\Gamma\check{D}(\Gamma) \otimes \check{S}^0 \otimes \check{T}^0$ and $\mathrm{X}\check{D}(\mathrm{X}) \otimes \check{S}^0 \otimes \check{T}^0$ types, where varied circles and segments represent oscillating electron densities and bond orders, respectively, while irregularly arranged circles denote lattice distortion. They consist of a paramagnetism (PM) of full symmetry, charge density waves (CDWs) within the metal or halogen sublattice, bond order waves (BOWs) along various paths, and x- or a-direction charge-polarized (CP) states. Metal (halogen)-sublattice CDW states are referred to as $(\sigma_1, \sigma_2, \sigma_3, \sigma_4)$, where the signatures $\sigma_l = \pm, 0$ denote the charges relative to M^{3+} (X^-) on adjacent metal (halogen) sites forming a square section of the square prism. Lattice-distorted BOWs are distinguishable in a similar way. The zigzag BOWs $\Gamma A_{2g} \otimes \check{S}^0 \otimes \check{T}^0$, $\Gamma A_{1u} \otimes \check{S}^0 \otimes \check{T}^0$, $\mathrm{X} A_{2g} \otimes \check{S}^0 \otimes \check{T}^0$, and $\mathrm{X} A_{1u} \otimes \check{S}^0 \otimes \check{T}^0$ are invariant under the four-fold rotation C_{4z}, whose bond patterns are depicted only on one side of the four, whereas the zigzag BOW $\Gamma B_{2u} \otimes \check{S}^0 \otimes \check{T}^0$ is invariant under the two-fold rotation C_{2z}, whose bond pattern is depicted only on two sides of the four, for the sake of readability.

abbreviated as M-CDW, X-CDW, and BOW, respectively. No lattice distortion accompanies BOW within the present Hamiltonian (1). Every BOW state may be stabilized by direct electron transfers on the oscillating bonds and their interactions with phonons. Lattice displacements perpendicular to the chain direction may stabilize ∧- and/or □-BOWs, but such observations have never been reported so far. We find twice four kinds of cell-doubled CDW states. There is an X version of every M-CDW. Although all the CDW states gain a condensation energy due to their Peierls distortion, they are not necessarily gapped. M- and X-CDW of the $(0+0-)$ type are *half metallic*, where two legs are valence-delocalized, while the rest are valence-trapped. Such states as cell-doubled but partially metallic are generally possible in tubed MX compounds, including triangular prisms [68], whose little groups $\mathbf{D}(\mathrm{X})$ all have a two-dimensional irreducible representation of axial isotropy subgroup. All the other CDW states are *fully gapped* in the center of the reduced Brillouin zone. Since the M d_{z^2} orbitals are half filled and lie higher in energy than the fully occupied X p_z orbitals, d-electron CDW states of π oscillation are most likely to appear in neutral samples at low temperatures.

3. Quantum Phase Competition

Now we calculate each of the HF energies individually and compare them. Six examples of the thus-obtained ground-state phase diagrams are shown in Fig. 3. While we have numerous parameters, previous theoretical and experimental studies on M family compounds serve as a reference. There is a strong resemblance between the platinum-halide prism [Pt(en)(bpy)I]$_4$(NO$_3$)$_8$ [66] and ladder (bpy)[Pt(dien)Br]$_2$Br$_4$ · 2H$_2$O [58] in particular. Referring to optical investigations [61, 63] of this ladder compound, we assume that $t_{MX}^{\parallel} = 1.46\,\mathrm{eV}$ and $t_{MM}^{\perp} = 0.47\,\mathrm{eV}$. Model studies [10, 11, 72], first-principle calculations [73], and photostructural investigations [25, 41, 74] on MX and MMX chains give further information about electron-electron and electron-lattice interactions. We adopt coupling and spring constants, $\alpha = 0.46\,\mathrm{eV}/\text{Å}$, $\beta_M = \beta_X = 2.08\,\mathrm{eV}/\text{Å}$, and $K_{MX}^{\parallel} = 8.00\,\mathrm{eV}/\text{Å}^2$. All other electronic parameters are adjusted to the convincing set: [66] $U_M = 1.68\,\mathrm{eV}$, $U_X = 1.24\,\mathrm{eV}$, $V_{MX}^{\parallel} = 0.61\,\mathrm{eV}$, $V_{MX}^{\times\!\times} = 0.31\,\mathrm{eV}$, $V_{MM}^{\parallel} = 0.35\,\mathrm{eV}$, $V_{MM}^{\perp} = 0.33\,\mathrm{eV}$, $V_{MM}^{\times\!\times} = 0.16\,\mathrm{eV}$, $V_{MM}^{\times} = 0.14\,\mathrm{eV}$, $V_{XX}^{\parallel} = 0.15\,\mathrm{eV}$, $V_{XX}^{\perp} = 0.10\,\mathrm{eV}$, $V_{XX}^{\times\!\times} = 0.07\,\mathrm{eV}$, $V_{XX}^{\times} = 0.03\,\mathrm{eV}$, and $\varepsilon_M - \varepsilon_X = 2.00\,\mathrm{eV}$. We refer to these parameters as the standard and use them, setting the band filling equal to three quarters, unless otherwise noted.

Figure 3(a) demonstrates quantum phase transitions with varying Coulomb interactions. The competition between M-CDW states is straightforwardly understandable when we assume the X p_z orbitals to be fully filled and thus inactive. Those of $(++++)$, $(+-+-)$ and $(++--)$ are stabilized with increasing $V_{MM}^{\times\!\times}$, $V_{MM}^{\perp}$, and $V_{MM}^{\times}$, respectively. On condition that $t_{MM}^{\perp} \ll t_{MX}^{\parallel} \ll \beta_M^2/K_{MX}^{\parallel}$, all the platinum ions are strongly valence-trapped and their d-electron energies read as

$$\frac{E_{(++++)}}{N} = 2\Big(U_M + 2\varepsilon_M + 4V_{MM}^{\perp} + 2V_{MM}^{\times} - \frac{2\beta_M^2}{K_{MX}^{\parallel}}\Big), \tag{15}$$

$$\frac{E_{(+-+-)}}{N} = 2\Big(U_M + 2\varepsilon_M + 2V_{MM}^{\times} + 8V_{MM}^{\times\times} - \frac{2\beta_M^2}{K_{MX}^{\parallel}}\Big), \tag{16}$$

$$\frac{E_{(++--)}}{N} = 2\Big(U_M + 2\varepsilon_M + 2V_{MM}^{\perp} + 4V_{MM}^{\times\times} - \frac{2\beta_M^2}{K_{MX}^{\parallel}}\Big), \tag{17}$$

for the three competing M-CDW states. Then the phase boundaries are given by $V_{MM}^{\perp} = 2V_{MM}^{\times\times} \pm V_{MM}^{\times}$ and we find $(++--)$-M-CDW stablest at the standard parameters ($\circ$). Therefore, we have to correct this naivest scenario taking account of the kinetic energies of d electrons. The leading valence-delocalization effect comes from intrachain transfers, but it equally acts on all in the decoupled-chain limit. The effective corrections originate from interchain transfers and read, disregarding $t_{MM}^{\parallel}$, as

$$\frac{\Delta E_{(+-+-)}^{\perp}}{N} = \frac{-8(t_{MM}^{\perp})^2}{4V_{MM}^{\parallel} + 3V_{MM}^{\perp} - 2V_{MM}^{\times} - 8V_{MM}^{\times\times} - U_M + 4\beta_M^2/K_{MX}^{\parallel}}, \tag{18}$$

$$\frac{\Delta E_{(++--)}^{\perp}}{N} = \frac{-4(t_{MM}^{\perp})^2}{4V_{MM}^{\parallel} - V_{MM}^{\perp} + 2V_{MM}^{\times} - U_M + 4\beta_M^2/K_{MX}^{\parallel}}. \tag{19}$$

$(++++)$-M-CDW gains no energy in this context, because neither of the constituent fully-occupied and totally-vacant plaquettes of platinum ions allows any valence delocalization. It is clear that $(+-+-)$-M-CDW gains the most from the interchain electron transfer: $0 = \Delta E_{(++++)}^{\perp} < \Delta E_{(++--)}^{\perp} < \Delta E_{(+-+-)}^{\perp}$. The thus-corrected phase boundaries are in excellent agreement with Figs. 3(a)–3(c). From the theoretical point of view, $(++--)$-M-CDW is characteristic of a tubed lattice in that it belongs to a two-dimensional representation. On the other hand, $(++++)$-M-CDW and $(+-+-)$-M-CDW have good analogy with CDW states of the IP and OP types, respectively, found in MX ladder compounds [61].

Figure 3(d) shows that the Peierls instability in a platinum-halide square-prism is conditional, where the electron-lattice coupling strength seriously affects the ground state. Unless α and/or β are sufficiently large, the ground state stays valence-delocalized. With increasing α and β, PM disappears into $(+-+-)$-M-CDW and it further turns into $(++--)$-M-CDW. A single MX chain is unconditionally distorted [21, 44], but coupled MX chains, whether tubed or not, are conditionally distorted [60, 67, 68]. It is the case with organic polymers as well. The Peierls instability in polyacetylene is unconditional, whereas those in polyacene are conditional [75, 76].

Now we discuss whether and how the novel phase, $(0+0-)$-M-CDW, appears in practice. While we have fully tuned both correlation parameters and coupling constants, we were not able to find its stabilization into the ground state as far as the band filling ν stays three quarters. However, there is a possibility of $(0+0-)$-M-CDW replacing $(++--)$-M-CDW under doping, which is demonstrated in Figs. 3(e) and 3(f). The system going

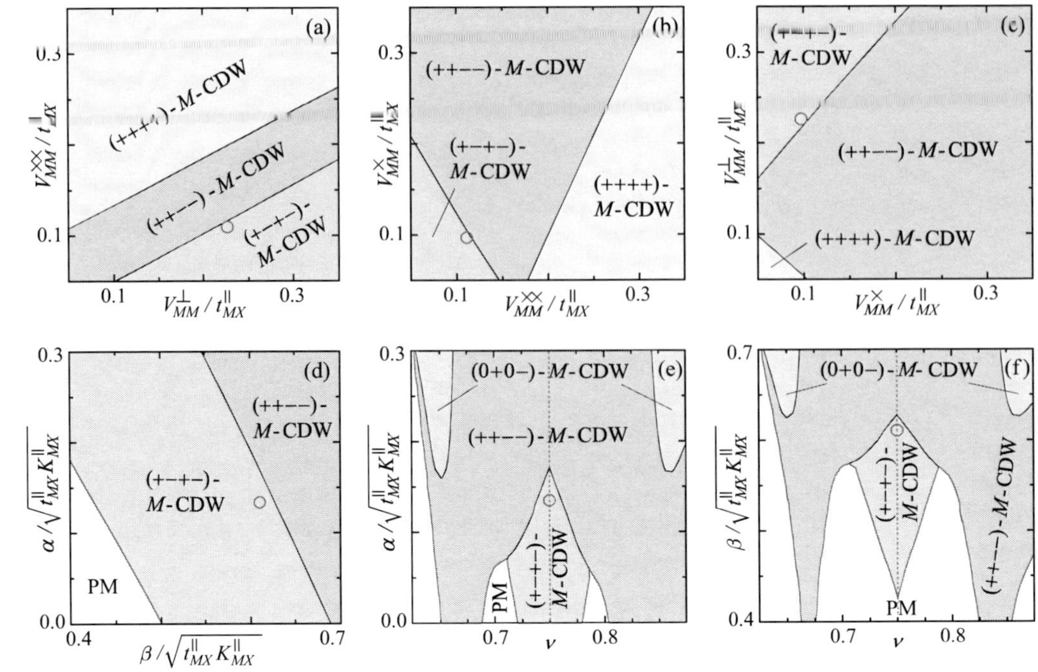

Figure 3. (Color online) Ground-state phase diagrams with varying Coulomb interactions [(a)–(c)] and those with varying coupling constants and band filling [(d)–(f)]. Phase boundaries drawn in black and red (or toned down in monochromes) denote transitions of the first and second order, respectively. ○ indicates the standard parametrization, while the dotted line separating the hole- and electron-doped regions, $\nu = 3/4$, is a guide for eyes.

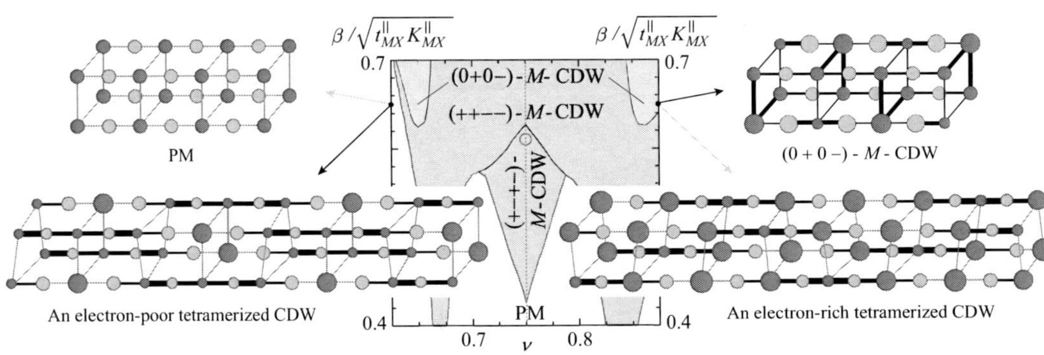

Figure 4. (Color online) Competing phases under doping. Two examples are given on the line of $\beta/\sqrt{t_{MX}^{\parallel}K_{MX}^{\parallel}} = 0.65$, $\nu = 0.625$ and $\nu = 0.875$, at the former of which a tetramerized CDW is stabler than PM, while at the latter of which $(0+0-)$-M-CDW indeed plays the ground state overcoming any other metastable states, including a tetramerized CDW.

away from $\nu = 3/4$ with somewhat larger α and/or β than the standard ones can encounter $(0+0-)$-M-CDW. There occurs a partially metallic state in between the totally valence-trapped and fully metallic states. Electrochemical doping, by exposing a single crystal to halogen vapor [77], for instance, possibly causes successive phase transitions towards the fully metallic state with conductivity increasing stepwise. In the wholly distorted M-CDWs, $(++++)$, $(+-+-)$, and $(++--)$, electrons are doped equally into electron-poor platinum sites, Pt^{4+} [$\mathrm{Pt}^{(3-\delta+\delta')+}$ $(0 \leq \delta < \delta' < 1)$ in general], whereas holes into electron-rich platinum sites, Pt^{2+} [$\mathrm{Pt}^{(3-\delta-\delta')+}$ $(0 \leq \delta < \delta' < 1)$ in general]. In $(0+0-)$-M-CDW, on the other hand, the metallic chains are doped first, while the Peierls-distorted chains remain three-quarter filled, or half-filled in the single-band picture, because the metallic Pt^{3+} bands are sandwiched between the bonding Pt^{2+} and antibonding Pt^{4+} bands far apart from them (see Fig. 5).

We should mention structural instabilities of longer period in this context. There may be CDW solutions of $0 < Q < \pi/c$ under doping. Figure 4 demonstrates such examples. At $\nu = 3/4 \pm 1/16$ and $\nu = 3/4 \pm 1/8$, a platinum-halide square prism can bc octamerized and tetramerized, respectively, to form commensurate CDWs. There exists a "heterogeneously" CDWd prism, as well as a wholly octamerized one, at $\nu = 3/4 \pm 1/16$, which consists of two $3/4$-filled dimerized and two $(3/4 \pm 1/8)$-filled tetramerized chains. Some of these solutions closely compete with but still stay higher in energy than dimerized M-CDWs, unless the coupling constants are sufficiently large. In the case that all the X solutions disappear into PM, long-period commensurate CDWs may be the most promising of all the broken-symmetry solutions. With increasing electron-lattice interactions, long-period ground states are possibly of occurrence even in neutral chains [78]. However, it may not be the case with our platinum-halide prisms, especially with iodo complexes.

4. Band Structure

Let us observe the electronic properties of various M-CDWs in more detail. We calculate their energy structures with the same set of parameters and show them in Fig. 5. $(++++)$-M-CDW (a), $(+-+-)$-M-CDW (b), and $(++--)$-M-CDW (c) are fully gapped, whereas $(0+0-)$-M-CDW (d) is gapless and metallic. The metallic bands are made predominantly from trivalent platinum ions [see $\rho_{n:lM}(\varepsilon)$ in Fig. 5(d)], but their splitting in two evidences interchain orbital hybridization. The wider one is purely composed of paramagnetic platinum ions, Pt^{3+} [$\mathrm{Pt}^{(3-\delta)+}$ $(0 \leq \delta < 1)$ in general], whereas the narrower one consists of those mixed with valence-trapped platinum ions, $\mathrm{Pt}^{(3\pm1)+}$ [$\mathrm{Pt}^{(3-\delta\pm\delta')+}$ $(0 \leq \delta < \delta' < 1)$ in general]. A possible reduction of the average valence is not due to the interchain d-electron transfer but comes from d-p hybridization. In $(0+0-)$-M-CDW, holes are doped into these metallic bands, that is, mainly into the two uniform chains, while the rest two remain three-quarter filled until the metallic chains are emptied. In the other wholly distorted M-CDWs, on the other hand, all the four chains equally contribute to the highest-lying filled band [see $\rho_{n:lM}(\varepsilon)$ in Figs. 5(a)–(c)] and therefore, doped holes immediately eat into the CDW structures. Thus and thus, there is a possibility of $(0+0-)$-M-CDW being stabilized into the ground state under moderate doping. Once the conduction electrons are removed out, $(0+0-)$-M-CDW is no more free from holes invading divalent platinum ions and suffers a rapid collapse.

Figure 5 further elucidate the optical features of M-CDWs. When the charge-transfer excitation energy is of eV order, the real part of the c-axis-polarized optical conductivity

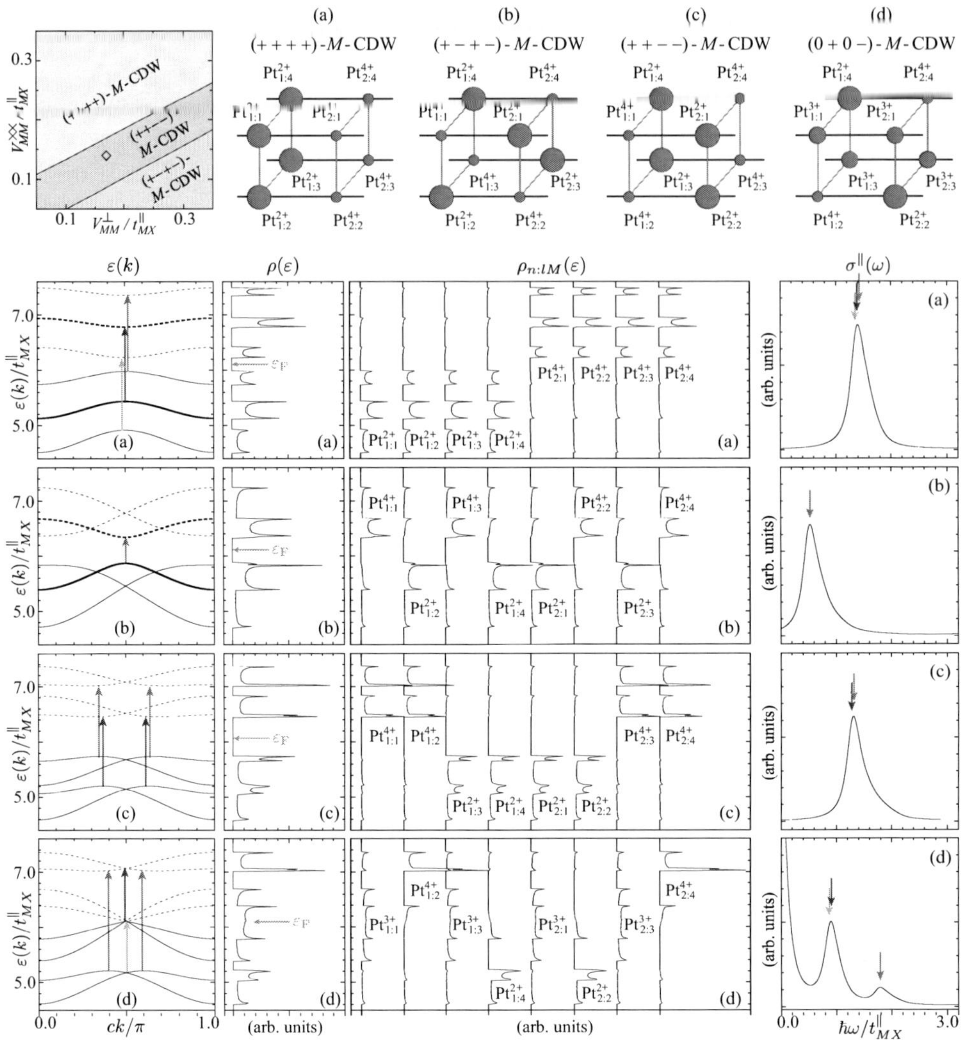

Figure 5. (Color online) Hartree-Fock calculations of the single-particle energy dispersion relation $\varepsilon(k)$, the local (total) density of states $\rho_{n:lA}(\varepsilon)$ $[\rho(\varepsilon) \equiv \sum_{n,l,A} \rho_{n:lA}(\varepsilon)]$, and the c-axis-polarized optical-conductivity spectrum $\sigma^{\parallel}(\omega)$ for M-CDW states of the $(++++)$ (a), $(+-+-)$ (b), $(++--)$ (c), and $(0+0-)$ (d) types under the standard parametrization but $V_{MM}^{\perp} = 0.25\,\text{eV}$ and $V_{MM}^{\times\times} = 0.20\,\text{eV}$ ($\diamond$ in the attached phase diagram), where all the four solutions are available. We focus on the highest-lying eight bands mainly of Pt character, where solid and dotted lines consist of occupied and vacant states, while thin and thick lines denote singly and doubly degenerate bands, respectively. $\varepsilon_{\rm F}$ indicates the Fermi energy. Major optical absorptions are indicated by vertical arrows.

reads as

$$\sigma_{\text{reg}}^{\parallel}(\omega) = \frac{\pi}{N\omega} \sum_{i \neq 0} |\langle E_i | \mathcal{J}^{\parallel} | E_0 \rangle|^2 \delta(E_i - E_0 - \hbar\omega), \tag{20}$$

where the relevant current operator $\mathcal{J}^{\parallel}$ is given by

$$\mathcal{J}^{\parallel} = \frac{iec}{2\hbar} \sum_{l=1}^{4} \sum_{n=1}^{N} \sum_{s=\uparrow,\downarrow} \{[t_{MX}^{\parallel} - \alpha(u_{n:lX}^{\parallel} - u_{n:lM}^{\parallel})] a_{n:lXs}^{\dagger} a_{n:lMs}$$

$$-[t_{MX}^{\parallel} - \alpha(u_{n+1:lM}^{\parallel} - u_{n:lX}^{\parallel})] a_{n+1:lMs}^{\dagger} a_{n:lXs} - \text{H.c.}\}, \tag{21}$$

while $|E_i\rangle$ is an arbitrary wave function of energy E_i ($E_0 \leq E_1 \leq E_2 \leq \cdots$). We define the ground state $|E_0\rangle$ as

$$|E_0\rangle = \prod_{\varepsilon_{\lambda s} \leq \varepsilon_{\text{F}}} a_{\lambda\uparrow}^{\dagger} a_{\lambda\downarrow}^{\dagger} |0\rangle, \tag{22}$$

where $|0\rangle$ is the true electron vacuum, ε_{F} is the Fermi energy, and $a_{\lambda s}^{\dagger}$ creates an electron of spin s in the HF eigenstate with an eigenvalue $\varepsilon_{\lambda s}$. Excited states can be calculated within and beyond the HF scheme, being generally defined as

$$|E_i\rangle = \sum_{\varepsilon_{\lambda s} \leq \varepsilon_{\text{F}} < \varepsilon_{\mu s}} f(\lambda, \mu, s; i) a_{\mu s}^{\dagger} a_{\lambda s} |E_0\rangle. \tag{23}$$

Every excited state of the HF type is a single Slater determinant, where $f(\lambda, \mu, s; i) = \delta_{\lambda\mu s, i}$. Those of the configuration-interaction (CI) type consist of resonating Slater determinants, where $f(\lambda, \mu, s; i)$ satisfies

$$\sum_{\varepsilon_{\lambda s} \leq \varepsilon_{\text{F}} < \varepsilon_{\mu s}} \langle E_0 | a_{\lambda' s'}^{\dagger} a_{\mu' s'} \mathcal{H} a_{\mu s}^{\dagger} a_{\lambda s} | E_0 \rangle f(\lambda, \mu, s; i) = E_i f(\lambda', \mu', s'; i), \tag{24}$$

that is, the unitary matrix $f(\lambda, \mu, s; i)$ diagonalizes the original Hamiltonian $\mathcal{H}$. Since the HF Hamiltonian $\mathcal{H}_{\text{HF}}$ is diagonal with respect to pure particle-hole states as

$$\langle E_0 | a_{\lambda' s'}^{\dagger} a_{\mu' s'} \mathcal{H}_{\text{HF}} a_{\mu s}^{\dagger} a_{\lambda s} | E_0 \rangle = \delta_{\lambda' \mu' s', \lambda \mu s} (E_0 - \varepsilon_{\lambda s} + \varepsilon_{\mu s}), \tag{25}$$

the residual component $\mathcal{H} - \mathcal{H}_{\text{HF}}$ mixes the Slater determinants and reduces the interband transition energy. While the optical conductivity for a fully gapped state may be evaluated in this manner, that for metallic states such as PM and $(0+0-)$-M-CDW demands further consideration to the low-frequency contribution, that is, the Drude spectral weight [79, 80],

$$\sigma_0^{\parallel} = \frac{\pi}{N} \left[\langle E_0 | (-\mathcal{T}^{\parallel}) | E_0 \rangle - 2 \sum_{i \neq 0} \frac{|\langle E_i | \mathcal{J}^{\parallel} | E_0 \rangle|^2}{E_i - E_0} \right], \tag{26}$$

with the kinetic-energy operator

$$\mathcal{T}^{\parallel} = \left(\frac{ec}{2\hbar} \right)^2 \sum_{l=1}^{4} \sum_{n=1}^{N} \sum_{s=\uparrow,\downarrow} \{[t_{MX}^{\parallel} - \alpha(u_{n+1:lM}^{\parallel} - u_{n:lX}^{\parallel})] a_{n+1:lMs}^{\dagger} a_{n:lXs}$$

$$-[t_{MX}^{\parallel} - \alpha(u_{n:lX}^{\parallel} - u_{n:lM}^{\parallel})] a_{n:lXs}^{\dagger} a_{n:lMs} + \text{H.c.}\}. \tag{27}$$

In an attempt to ascertain which excitation is contributive to optical absorption, we calculate the optical-conductivity spectra within the HF scheme. Those of the wholly distorted M-CDWs, $\sigma^{\|}(\omega) = \sigma^{\|}_{\mathrm{reg}}(\omega)$, and that of the half metallic M CDW, $\sigma^{\|}(\omega) = \sigma^{\|}_{0}\delta(\omega) + \sigma^{\|}_{\mathrm{reg}}(\omega)$, are shown in Figs. 5(a)–5(c) and in Fig. 5(d), respectively, where every spectral line is Lorentzian-broadened by a width of $0.12t^{\|}_{MX}$ and higher-energy absorptions in which $X\,p_z$ orbitals are involved are omitted. There is (are) only one, two, and three type(s) of contributive interband electronic transitions to the optical absorption in $(+-+-)$-M-CDW, $(++--)$-M-CDW, and $(++++)$-M-CDW, respectively. Their spectra are all shaped alike under the present parametrization, but the last two are possibly double- and triple-peaked, especially in the d-p-π description [61]. The spectrum of $(0+0-)$-M-CDW is indeed multifeatured. The highest-lying band originates within valence-trapped platinum ions, whereas the zero-frequency spectral weight comes predominantly from valence-delocalized platinum ions. All the platinum ions are involved in the mid band. Both continuous and discontinuous metal-insulator transitions may be of occurrence in square-prism-shaped platinum-chloride complexes and the Drude weight serves as a direct and sensitive measure of them.

5. Optical Observations

It is highly interesting to apply our theory to the motive material $[Pt(en)(bpy)I]_4(NO_3)_8 \cdot 16H_2O$. The c-axis-polarized optical-conductivity spectrum of this compound is characterized by two overlapping but well-distinguishable absorption bands, which sit half an eV or more apart from each other [66]. Such a well-split, double-peaked intervalence charge-transfer absorption spectrum is hardly understandable within any of $(++++)$-M-CDW, $(+-+-)$-M-CDW, and $(++--)$-M-CDW, which are likely states in consideration of the single-crystal Raman spectrum [66]. We are thus led to examine a mixed ground state on the assumption of the material sitting in close vicinity of a phase boundary. All the Peierls instabilities continuously arise from the valence-delocalized state of full symmetry, whereas any quantum phase transition between them is discontinuous. Therefore, two or more mixed-valent states of distinct symmetry possibly coexist in the ground state. We inquire which can be coexistent with which. "$(+-+-)$-M-CDW with $(++--)$-M-CDW or $(++--)$-M-CDW with $(++++)$-M-CDW," answer Figs. 3(a)–3(c).

We consider both cases in Fig. 6. The optical-conductivity spectra of a trial ground state and the coexistent metastable state, together with the two-to-one mixture of their intensities, are shown in comparison with experimental findings [66]. Here we calculate them beyond the HF approximation [47, 81], that is, taking account of configuration interactions between the HF Slater determinants up to the single-excitation level [61, 82]. The mixed ground state of the $(+-+-)$ and $(++--)$ types reproduces the observations well, whereas that of the $(++--)$ and $(++++)$ types does not work well. The absorption peaks of $(+-+-)$-M-CDW and $(++--)$-M-CDW appear well apart from each other, while those of $(++--)$-M-CDW and $(++++)$-M-CDW sit close together, no matter how we tune parameters. We should be reminded in this context that the highest-occupied-molecular-orbital (HOMO)-to-lowest-unoccupied-molecular-orbital (LUMO) transition through a c-axis-polarized photon is allowed in $(+-+-)$-M-CDW but forbidden in $(++--)$-M-CDW and $(++++)$-

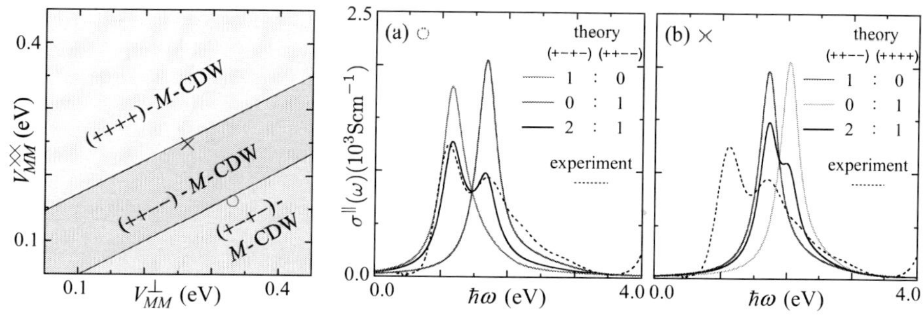

Figure 6. (Color online) Single-excitation configuration-interaction calculations (solid lines) of the c-axis-polarized optical-conductivity spectrum in comparison with experimental findings [66] (dotted lines). We consider superposing two coexistent M-CDW states in close vicinity of their phase boundary: (a) The standard parametrization (○); (b) that but $V_{MM}^{\perp} = 0.26\,\mathrm{eV}$ and $V_{MM}^{\times\times} = 0.25\,\mathrm{eV}(\times)$.

M-CDW. Therefore, the absorption spectrum of the first state is maximal at the Peierls-gap energy, whereas those of the last two states are both structured at much higher energies. A diffuse X-ray scattering experiment [66] also supports this scenario.

Closing Remarks

The novel nanotubular assembly $[\mathrm{Pt(en)(bpy)I}]_4(\mathrm{NO_3})_8 \cdot 16\mathrm{H_2O}$ was designed by a bottom-up process based on metal-organic building blocks. The bridging halide ions and the guest molecules can readily be tuned in an attempt to control the band gap. The intervalence charge-transfer energy, roughly proportional to $\beta_M/\sqrt{t_{MX}^{\|}K_{MX}^{\|}}$, indeed increases in the order I $<$ Br $<$ Cl, whereas it slightly decreases with H_2O away. Ni substitution for Pt will stimulate the brandnew interest in novel spin arrangements, bringing d electrons into stronger correlation. Mixed-metal compounds might demonstrate a competition [83] between Peierls and Mott insulators full of variety.

A recent attempt to fabricate an MX triangular prism [68] has brought us a platinum-bromide triple helix, $[\mathrm{Pt(dien)Br}]_3(\mathrm{tpt})\mathrm{Br}_3(\mathrm{ClO}_4)_3$ [tpt $=$ $2,4,6$-tri(4-pyridyl)-$1,3,5$-triazine $=$ $\mathrm{C_{18}H_{12}N_6}$] [84]. The newly synthisized square-prism-shaped iodo platinum complexes are quite different from other nanotubes such as hydrocarbons [85] in terms of structural and electronic tunabilities. We are thus motivated into an elaborate theory and we further hope that such a study will stimulate growing chemical explorations.

Acknowledgments

We express special thanks to M. Ozaki for helpful comments on our calculation and to K. Otubo and H. Kitagawa for valuable information on their wonderful products.

A. Momentum Representation of the Hamiltonian

Equations (5) and (6) read as follows:

$$\langle k+Q:lX|t|k:lM\rangle = 2i\delta_{Q\Gamma}t^{\parallel}_{MX}\sin\frac{kc}{2} + \frac{2\alpha}{\sqrt{N}}\left[u^{\parallel}_{Q:lX}\cos\frac{kc}{2} - u^{\parallel}_{Q:lM}\cos\frac{(k+Q)c}{2}\right], \tag{28}$$

$$\langle k+Q:lM|t|k:lX\rangle = -2i\delta_{Q\Gamma}t^{\parallel}_{MX}\sin\frac{kc}{2} + \frac{2\alpha}{\sqrt{N}}\left[u^{\parallel}_{Q:lX}\cos\frac{(k+Q)c}{2} - u^{\parallel}_{Q:lM}\cos\frac{kc}{2}\right], \tag{29}$$

$$\langle k+Q:l+1M|t|k:lM\rangle = \langle k+Q:lM|t|k:l+1M\rangle = -\delta_{Q\Gamma}t^{\perp}_{MM}, \tag{30}$$

$$\langle k+Q:lM|t|k:lM\rangle = \delta_{Q\Gamma}\varepsilon_M - \frac{2i\beta_M}{\sqrt{N}}u^{\parallel}_{Q:lX}\sin\frac{Qc}{2}, \tag{31}$$

$$\langle k+Q:lX|t|k:lX\rangle = \delta_{Q\Gamma}\varepsilon_X - \frac{2i\beta_X}{\sqrt{N}}u^{\parallel}_{Q:lM}\sin\frac{Qc}{2}, \tag{32}$$

$$\langle k+Q:lM;k':lM|v|k:lM;k'+Q:lM\rangle = \frac{U_M}{N} + \frac{2V^{\parallel}_{MM}}{N}\cos Qc, \tag{33}$$

$$\langle k+Q:lX;k':lX|v|k:lX;k'+Q:lX\rangle = \frac{U_X}{N} + \frac{2V^{\parallel}_{XX}}{N}\cos Qc, \tag{34}$$

$$\langle k+Q:lM;k':lX|v|k:lM;k'+Q:lX\rangle = \langle k+Q:lX;k':lM|v|k:lX;k'+Q:lM\rangle = \frac{2V^{\parallel}_{MX}}{N}\cos\frac{Qc}{2}, \tag{35}$$

$$\langle k+Q:lM;k':l+1M|v|k:lM;k'+Q:l+1M\rangle = \langle k+Q:l+1M;k':lM|v|k:l+1M;k'+Q:lM\rangle = \frac{V^{\perp}_{MM}}{N} + \frac{2V^{\times\!\times}_{MM}}{N}\cos Qc, \tag{36}$$

$$\langle k+Q:lX;k':l+1X|v|k:lX;k'+Q:l+1X\rangle = \langle k+Q:l+1X;k':lX|v|k:l+1X;k'+Q:lX\rangle = \frac{V^{\perp}_{XX}}{N} + \frac{2V^{\times\!\times}_{XX}}{N}\cos Qc, \tag{37}$$

$$\langle k+Q:lM;k':l+2M|v|k:lM;k'+Q:l+2M\rangle = \langle k+Q:l+2M;k':lM|v|k:l+2M;k'+Q:lM\rangle = \frac{V^{\times}_{MM}}{2N}, \tag{38}$$

$$\langle k+Q:lX;k':l+2X|v|k:lX;k'+Q:l+2X\rangle = \langle k+Q:l+2X;k':lX|v|k:l+2X;k'+Q:lX\rangle = \frac{V^{\times}_{XX}}{2N}, \tag{39}$$

$$\langle k+Q:lM;k':l+1X|v|k:lM;k'+Q:l+1X\rangle = \langle k+Q:l+1X;k':lM|v|k:l+1X;k'+Q:lM\rangle = \langle k+Q:l+1M;k':lX|v|k:l+1M;k'+Q:lX\rangle = \langle k+Q:lX;k':l+1M|v|k:lX;k'+Q:l+1M\rangle = \frac{2V^{\times}_{MX}}{N}\cos\frac{Qc}{2}. \tag{40}$$

B. Point Groups Relevant to Square-Prism-Shaped MX Compounds

Relevant subgroups of $\mathbf{D}_{4h}$ are given as follows:

$$\mathbf{C}_{2ah} = \mathbf{C}_{2a} + I\mathbf{C}_{2a}, \tag{41}$$
$$\mathbf{C}_{2xh} = \mathbf{C}_{2x} + I\mathbf{C}_{2x}, \tag{42}$$
$$\mathbf{C}_{2av} = \mathbf{C}_{2a} + IC_{2b}\mathbf{C}_{2a}, \tag{43}$$
$$\mathbf{C}_{2xv} = \mathbf{C}_{2x} + IC_{2y}\mathbf{C}_{2x}, \tag{44}$$
$$\mathbf{D}_{2ah} = \mathbf{D}_{2a} + I\mathbf{D}_{2a}, \tag{45}$$
$$\mathbf{D}_{2h} = \mathbf{D}_{2} + I\mathbf{D}_{2}, \tag{46}$$
$$\mathbf{D}_{2ad} = \mathbf{D}_{2a} + IC_{2x}\mathbf{D}_{2a}, \tag{47}$$
$$\mathbf{D}_{2d} = \mathbf{D}_{2} + IC_{2a}\mathbf{D}_{2}, \tag{48}$$
$$\mathbf{C}_{4h} = \mathbf{C}_{4} + I\mathbf{C}_{4}, \tag{49}$$
$$\mathbf{C}_{4v} = \mathbf{C}_{4} + IC_{2x}\mathbf{C}_{4}, \tag{50}$$
$$\mathbf{D}_{4} = \mathbf{C}_{4} + C_{2x}\mathbf{C}_{4}, \tag{51}$$
$$\mathbf{D}_{4h} = \mathbf{D}_{4} + I\mathbf{D}_{4}, \tag{52}$$

with

$$\mathbf{C}_{2a} = \{E, C_{2a}\}, \tag{53}$$
$$\mathbf{C}_{2x} = \{E, C_{2x}\}, \tag{54}$$
$$\mathbf{D}_{2a} = \{E, C_{2z}, C_{2a}, C_{2b}\} = \mathbf{C}_{2} + C_{2a}\mathbf{C}_{2}, \tag{55}$$
$$\mathbf{D}_{2} = \{E, C_{2z}, C_{2x}, C_{2y}\} = \mathbf{C}_{2} + C_{2x}\mathbf{C}_{2}, \tag{56}$$
$$\mathbf{C}_{4} = \{E, C_{4z}, C_{2z}, C_{4z}^{-1}\}. \tag{57}$$

References

[1] M. B. Robin and P. Day, in *Advances in Inorganic Chemistry and Radiochemistry*, edited by H. J. Emeleus (Academic, New York, 1967), Vol. 10, p. 247.

[2] P. Day, in *Low Dimensional Cooperative Phenomena*, edited by H. J. Keller (Plenum, New York, 1974), p.191.

[3] H. J. Keller, in *Extended Linear Chain Compounds*, edited by J. S. Miller (Plenum, New York, 1982), p. 357.

[4] D. S. Martin Jr., in *Extended Linear Chain Compounds*, edited by J. S. Miller (Plenum, New York, 1982), p. 434.

[5] R. J. H. Clark, in *Advances in Infrared and Raman Spectroscopy*, edited by R. J. H. Clark and R. E. Hester (Wiley Heyden, New York, 1984) Vol. 11, p. 95.

[6] M.-H. Whangbo, in *Crystal Chemistry and Properties of Materials with Quasi-One-Dimensional Structures*, edited by J. Rouxel (Reidel, New York, 1986), p. 27.

[7] K. Nasu, *J. Phys. Soc. Jpn.* 52, 3865 (1983).

[8] K. Nasu, *J. Phys. Soc. Jpn.* 53, 302 (1984).

[9] K. Nasu, *J. Phys. Soc. Jpn.* 53, 427 (1984).

[10] J. T. Gammel, A. Saxena, I. Batistić, A. R. Bishop, and S. R. Phillpot, *Phys. Rev. B* 45, 6408 (1992).

[11] S. M. Weber-Milbrodt, J. T. Gammel, A. R. Bishop, and E. Y. Loh, Jr., Phys. Rev. B 45, 6435 (1992).

[12] H. Wolfram, Dissertation, Königsberg (1900).

[13] H. Reihlen and E. Flohr, *Ber. Dtsch. Chem. Ges.* 67, 2010 (1934).

[14] B. M. Craven and D. Hall, *Acta Cryst.* 14, 475 (1961).

[15] M. Tanaka, S. Kurita, Y. Okada, T. Kojima, and Y. Yamada, *Chem. Phys.* 96, 343 (1985).

[16] R. Ikeda, T. Tamura, and M. Yamashita, *Chem. Phys. Lett.* 173, 466 (1990).

[17] H. Toftlund and O. Simonsen, *Inorg. Chem.* 23, 4261 (1984).

[18] K. Toriumi, Y. Wada, T. Mitani, S. Bandow, M. Yamashita, and Y. Fujii, *J. Am. Chem. Soc.* 111, 2341 (1989).

[19] V. Robert, S. A. Borshch, and B. Bigot, *Inorg. Chem.* 35, 3913 (1996).

[20] M. Alouani, J. W. Wilkins, R. C. Albers, and J. M. Wills, *Phys. Rev. Lett.* 71, 1415 (1993).

[21] D. Baeriswyl and A. R. Bishop, *J. Phys. C* 21, 339 (1988).

[22] A. Mishima and K. Nasu, *Phys. Rev. B* 39, 5758 (1989).

[23] A. Mishima and K. Nasu, Phys. Rev. B 39, 5763 (1989).

[24] N. Kuroda, M. Sakai, Y. Nishina, M. Tanaka, and S. Kurita, *Phys. Rev. Lett.* 58, 2122 (1987).

[25] S. Kurita, M. Haruki, and K. Miyagawa, *J. Phys. Soc. Jpn.* 57, 1789 (1988).

[26] R. J. Donohoe, S. A. Ekberg, C. D. Tait, and B. I. Swanson, *Solid State Commun.* 71, 49 (1989).

[27] H. Okamoto and M. Yamashita, *Bull. Chem. Soc. Jpn.* 71, 2023 (1998).

[28] S. Yamamoto, *Phys. Lett. A* 258, 183 (1999).

[29] S. Yamamoto, *Phys. Rev. B* 63, 125124 (2001).

[30] M. Kuwabara and K. Yonemitsu, *J. Mater. Chem.* 11, 2163 (2001).

[31] M. Yamashita, S. Takaishi, A. Kobayashi, H. Kitagawa, H. Matsuzaki, and H. Okamoto, *Coord. Chem. Rev.* 250, 2335 (2006).

[32] C.-M. Che, F. H. Herbstein, W. P. Schaefer, R. E. Marsh, and H. B. Gray, *J. Am. Chem. Soc.* 105, 4604 (1983).

[33] M. Kurmoo and R. J. H. Clark, *Inorg. Chem.* 24, 4420 (1985).

[34] C. Bellitto, A. Flamini, L. Gastaldi, and L. Scaramuzza, *Inorg. Chem.* 22, 444 (1983).

[35] H. Kitagawa, N. Onodera, T. Sonoyama, M. Yamamoto, T. Fukawa, T. Mitani, M. Seto and, Y. Maeda, *J. Am. Chem. Soc.* 121, 10068 (1999).

[36] S. A. Borshch, V. Robert, and B. Bigot, *J. Chem. Phys.* 109, 4562 (1998).

[37] S. Yamamoto, *J. Phys. Soc. Jpn.* 70, 1198 (2001).

[38] B. I. Swanson, M. A. Stroud, S. D. Conradson, and M. H. Zietlow, *Solid State Commun.* 65, 1405 (1988).

[39] S. Yamamoto, *Phys. Rev. B* 64, 140102(R) (2001).

[40] S. Yamamoto, *J. Phys. Chem. Solids* 63, 1489 (2002).

[41] H. Matsuzaki, T. Matsuoka, H. Kishida, K. Takizawa, H. Miyasaka, K. Sugiura, M. Yamashita and, H. Okamoto, *Phys. Rev. Lett.* 90, 046401 (2003).

[42] K. Yonemitsu and N. Miyashita, *Phys. Rev. B* 68, 075113 (2003).

[43] S. Yamamoto and M. Ichioka, *J. Phys. Soc. Jpn.* 71, 189 (2002).

[44] S. Yamamoto, *Phys. Rev. B* 66, 165113 (2002).

[45] A. Kobayashi, S. Kitao, M. Seto, R. Ikeda, and H. Kitagawa, *Inorg. Chem.* 48, 8044 (2009).

[46] M. Mitsumi, T. Murase, H. Kishida, T. Yoshinari, Y. Ozawa, K. Toriumi, T. Sonoyama, H. Kitagawa, and T. Mitani, *J. Am. Chem. Soc.* 123, 11179 (2001).

[47] J. Ohara and S. Yamamoto, *Phys. Rev. B* 73, 045122 (2006).

[48] S. Yamamoto and J. Ohara, *J. Phys.: Condens. Matter* 20, 415215 (2008).

[49] H. Tanaka, S. Kuroda, T. Yamashita, M. Mitsumi, and K. Toriumi, *Phys. Rev. B* 73, 245102 (2006).

[50] H. Tanaka, H. Nishiyama, S. Kuroda, T. Yamashita, M. Mitsumi, and K. Toriumi, *Phys. Rev. B* 78, 033104 (2008).

[51] C. Bellitto, G. Dessy, and V. Fares, *Inorg. Chem.* 24, 2815 (1985).

[52] S. Nakano, Y. Kitagawa, T. Kawakami, and K. Yamaguchi, *Polyhedron* 22, 2027 (2003).

[53] J. Ohara and S. Yamamoto, *J. Phys. Chem. Solids* 66, 1571 (2005).

[54] S. Ikeuchi, Y. Yamamura, Y. Yoshida, M. Mitsumi, K. Toriumi, and K. Saito, *J. Phys. Soc. Jpn.* 78, 094704 (2009).

[55] M. Mitsumi, Y. Yoshida, A. Kohyama, Y. Kitagawa, Y. Ozawa, M. Kobayashi, K. Toriumi, M. Tadokoro, N. Ikeda, M. Okumura, and M. Kurmoo, *Inorg. Chem.* 48, 6680 (2009).

[56] K. Sakai, Y. Tanaka, Y. Tsuchiya, K. Hirata, T. Tsubomura, S. Iijima, and A. Bhattacharjee, *J. Am. Chem. Soc.* 120, 8366 (1998).

[57] J. Yi, T. Miyabayashi, M. Ohashi, T. Yamagata, and K. Mashima, *Inorg. Chem.* 43, 6596 (2004).

[58] A. Kobayashi and H. Kitagawa *J. Am. Chem. Soc.* 128, 12066 (2006).

[59] D. Kawakami, M. Yamashita, S. Matsunaga, S. Takaishi, T. Kajiwara, H. Miyasaka, K. Sugiura, H. Matsuzaki, and H. Okamoto, *Angew. Chem.* 118, 7372 (2006).

[60] K. Funase and S. Yamamoto, *J. Phys. Soc. Jpn.* 75, 044717 (2006).

[61] S. Yamamoto and J. Ohara, *Phys. Rev. B* 76, 235116 (2007).

[62] K. Iwano and Y. Shimoi, *J. Phys. Soc. Jpn.* 76, 063708 (2007).

[63] S. Yamamoto and J. Ohara, *J. Mater. Sci.: Mater. Electron.* 20, S367 (2009).

[64] T. Suemoto, H. Nakao, M. Nakajima, and H. Kitagawa, *J. Chem. Phys.* 134, 224503 (2011).

[65] H. Tanino, K. Kobayashi, *J. Phys. Soc. Jpn.* 52, 1446 (1983).

[66] K. Otsubo, Y. Wakabayashi, J. Ohara, S. Yamamoto, H. Matsuzaki, H. Okamoto, K. Nitta, T. Uruga, and H. Kitagawa, *Nature Mater.* 10, 291 (2011).

[67] J. Ohara and S. Yamamoto, *Europhys. Lett.* 87, 17006 (2009).

[68] S. Yamamoto, J. Ohara, and M. Ozaki, *J. Phys. Soc. Jpn.* 79, 044709 (2010).

[69] M. Ozaki, *Int. J. Quantum Chem.* 42, 55 (1992).

[70] S. Yamamoto and M. Ozaki, *Int. J. Quantum Chem.* 44, 949 (1992).

[71] M. Golubittsky, I. Stewart, and D. G. Schaeffer, *Singurality and Groups in Bifurcation Theory* (Springer-Verlag, Berlin, 1985), Vol. II, p. 83.

[72] S. D. Conradson, M. A. Stroud, M. H. Zietlow, B. I. Swanson, D. Baeriswyl, and A. R. Bishop, *Solid State Commun.* 65, 823 (1988).

[73] R. C. Albers, M. Alouani, J. M. Wills, and M. Springborg, *Synth. Met.* 42, 2739 (1991).

[74] L. Degiorgi, P. Wachter, M. Haruki, and S. Kurita, *Phys. Rev. B* 40, 3285 (1989).

[75] S. Yamamoto, *Phys. Rev. B* 78, 235205 (2008).

[76] L. Salem and H. C. Longuet-Higgins, *Proc. Roy. Soc. A* 255, 435 (1960).

[77] M. Haruki and S. Kurita, *Phys. Rev. B* 39, 5706 (1989).

[78] I. Batistić, J. T. Gammel, and A. R. Bishop, *Phys. Rev. B* 44, 13228 (1991).

[79] J. Wagner, W. Hanke, and D. J. Scalapino, *Phys. Rev. B* 43, 10517 (1991).

[80] D. J. Scalapino, S. R. White, and S. C. Zhang, *Phys. Rev. Lett.* 68, 2830 (1992).

[81] J. Ohara and S. Yamamoto, *Phys. Rev. B* 70, 115112 (2004).

[82] J. Ohara and S. Yamamoto, *J. Phys. Soc. Jpn.* 74, 250 (2005).

[83] H. Matsuzaki, K. Iwano, T. Aizawa, M. Ono, H. Kishida, M. Yamashita, and H. Okamoto, *Phys. Rev. B* 70, 035204 (2004).

[84] K. Otsubo and H. Kitagawa, private communication.

[85] S. Iijima, *Nature* 354, 56 (1991).

INDEX

A

B

C

D

E

F

G

H

I

J

K

L

M

N

O

P

Q

R

S

T

U

V

W

X

Y

Z